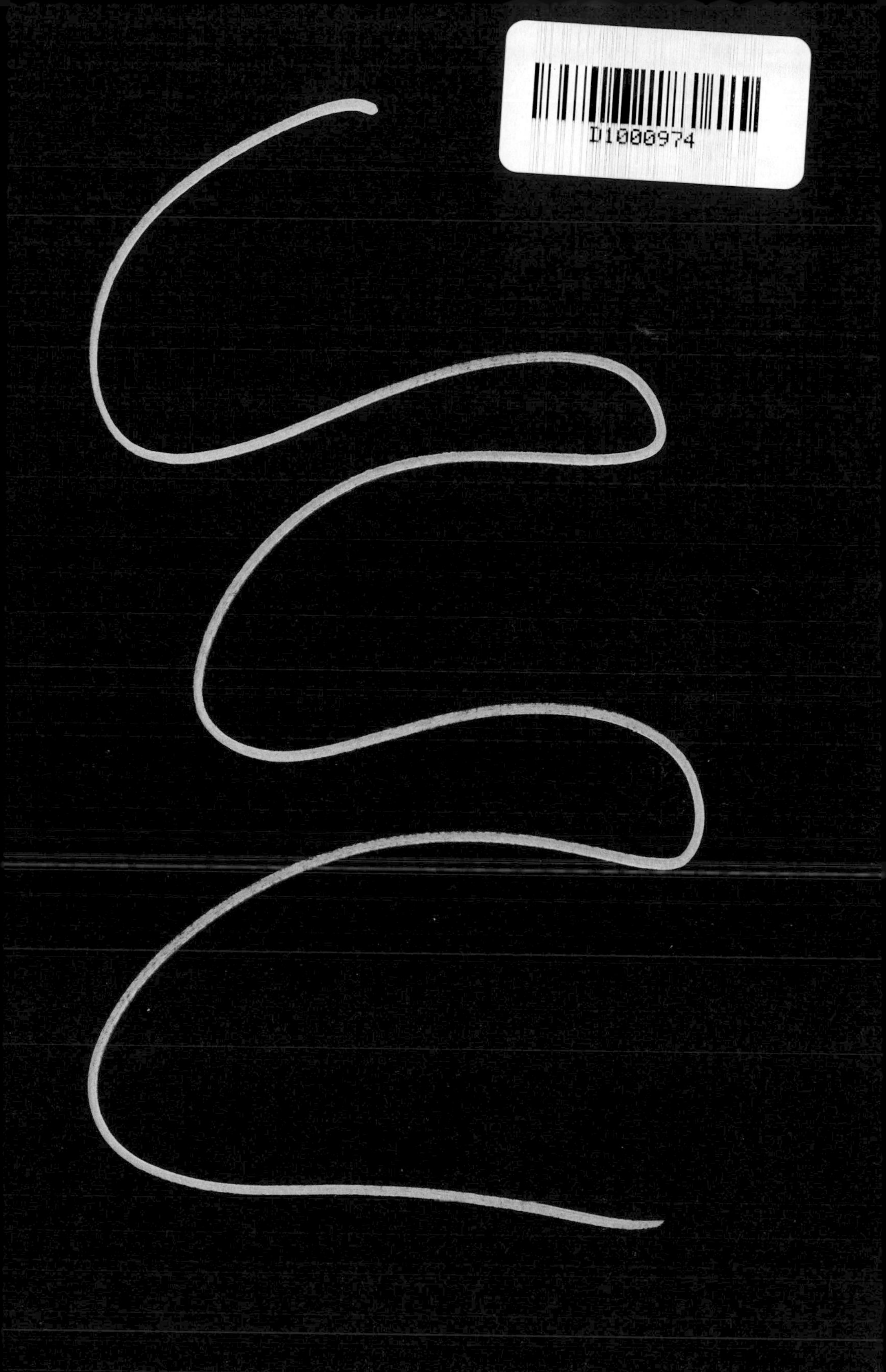

Biomedical Applications of Heat and Mass Transfer

Biomedical Applications of Heat and Mass Transfer

R. C. SEAGRAVE

The Iowa State University Press, Ames, Iowa

Richard C. Seagrave is a Professor at Iowa State University with a split appointment in chemical engineering and biomedical engineering. In addition to teaching courses in both programs, he directs research in chemical reactor analysis, transport phenomena in flow systems, and the simulation of physiological systems. He completed his undergraduate work at the University of Rhode Island and holds the Ph.D. degree in chemical engineering from Iowa State.

Composed and printed in the U.S.A.
First edition, 1971
International Standard Book Number: 0-8138-0195-8
Library of Congress Catalog Card Number: 71-146930

Contents

Preface

This text, which evolved from a one-quarter first-year graduate course constructed by the author, provides an introduction to the basic principles of chemical thermodynamics, energy transport, and mass transfer for students in the life sciences and for engineers working in areas of biomedical engineering. It is designed also to offer working examples of the applications of these topics in the study of physiology and in the design and operation of artificial organs and life support systems. It is hoped that the examples will strengthen the basic understanding of those already well versed in the study of energy and mass transfer and that those expert in the life sciences will be able to apply this competence to the physical principles presented.

The level of the text is suitable for advanced undergraduates and beginning graduate students. The subject matter presumes some competence in calculus and beginning differential equations, although advanced training in the latter, while helpful, is not necessary. It also presumes a reasonable background in chemistry and physics, with a physical chemistry sequence perhaps providing the ideal background. Finally, it is certainly desirable that the reader have had some exposure to a basic physiology sequence, although much of the material on the transfer processes can stand on its own.

As in any elementary interdisciplinary text, a certain amount of simplification must be tolerated on pedagogical grounds. It is hoped that the distillations encountered in this text, not only in the physiological realm but also in the physical-chemical area, are not excessive to the point of error, but serve primarily to elucidate the basic principles involved. More rigorous and advanced texts are certainly available in many of the areas covered here, and many of the topics are the subject of prodigious current research interest. This work can provide a much needed mechanism for relating these important areas at the beginning level, while providing the research worker with a starting point for more intensive study.

While the bibliography is organized on a chapter-by-chapter basis, it should be realized that many of the listings, in particular the textbooks, have more general applicability in providing both background and extension for this text. It should also be realized that the list is by no means exhaustive; that is, works selected are typical examples only.

In addition, many of the examples and some of the fundamental data used were selected from the listed works. The author has attempted to give proper credit in each section for these extractions and hopes the reader will realize that many of the examples are of such general nature that full and proper credit to the original source is extremely difficult.

The author is indebted to the classes of graduate students who patiently and scholastically provided the necessary feedback over several iterations of this material to bring it to its present form.

Examples

Notation

While it should be realized that most symbols in this text are defined in or near the section where they are employed and should be interpreted therewith, the symbols, subscripts, and superscripts most frequently used are listed below. As in any scientific text it is practically impossible to use dictinctly different symbols for each physical or chemical quantity and still employ the more universally accepted forms for many important variables. In this text, unique in the situation that two major disciplines (chemical engineering and physiology) are mixed, the compromise has been on the side of the latter goal; therefore the reader must bear some repetition, although, it is hoped, a minimum.

SYMBOL	*DEFINITION*
A	area (L^2)
a	acceleration (L/t^2)
a	activity
C	concentration (moles/L^3)
c	speed of light (L/t)
C	heat capacity, (Q/Mt)
D	diameter, (L)
d	derivative operator
$\mathfrak{D}$	diffusion coefficient, (L^2/t)
F	force, (F)
F	view factor
$\mathfrak{F}$	Faraday's constant
F	relative concentration
g	acceleration of gravity, (L/t^2)
g_c	conversion factor, (ML^2/Ft)
G	free energy
H	enthalpy

SYMBOL	*DEFINITION*
H	hematocrit
h	heat transfer coefficient
j	mass transfer flux, (M/L^2t)
J	mass transfer flux, (moles/L^2t)
k	mass transfer coefficient
K	overall mass transfer coefficient
K	permeability
K	kinetic energy
K	chemical equilibrium constant
k	thermal conductivity
i	summation index
j	summation index
L	length (L)
M	molecular weight
M	mass (M)
m	mass
$\dot{M}$	metabolic energy conversion rate
n	number of moles
n_i	mass transfer flux, (M/L^2t)
N	mass transfer flux, (moles/L^2t)
P	pressure
p	partial pressure
Q	volumetric flow rate, (L^3/t)
Q	heat (Q)
$\dot{Q}$	rate of heat transfer, (Q/t)
q	heat transfer flux, (Q/L^2t)
R	gas constant (PV/nT)
R	respiratory quotient
$\dot{R}$	rate of irreversible energy loss
r_i	chemical reaction rate, (M/L^3t)
R	radius
R_i	chemical reaction rate, (moles/L^3t)
S	entropy
t	time, (t)
T	temperature, (T)
$\dot{T}$	rate of sensible heat change
U	internal energy
v	velocity, (L/t)
V	volume, (L^3)
w	mass flow rate, (M/t)
W	weight (Mg)

SYMBOL	*DEFINITION*
W	work, (FL)
$\dot{W}$	rate of work, (FL/t)
x	distance, (L)
x	mole fraction, liquid
X	mole ratio, liquid
y	mole fraction, gas
Y	mole fraction ratio, gas
z	distance, (L)
r	radial distance, (L)

GREEK

β	coefficient of volume expansion
Δ	change
∂	partial derivative operator
δ	discrete change
ϵ	potential (electrical)
μ	fluid viscosity
π	hydrostatic pressure
π	3.1416
ρ	mass density
σ	Stefan-Boltzmann constant
λ	wavelength
ν	frequency
ϕ	potential energy
ϕ	distribution coefficient
Σ	sum
Ω	mass ratio

SUPERSCRIPTS

$\cdot$	rate or time derivative
$\hat{}$	per unit mass
$\sim$	per mole
o	standard state
$^{\circ}$(as in $^{\circ}$F)	degrees temperature
$*$	molar reference frame

SUBSCRIPTS

a, b, c, d	components A, B, C, D
A	ambient
A	alveolar

SYMBOL	*DEFINITION*
a	arterial
B	blood
c	cold
D	dead space
D	dialysate
c	convection
E	expired
f	film
f	food
g	gas
f	formation
h	hot
i	inside
i	species i
l	liquid
LM	logarithmic mean
m	muscle
max	maximum
min	minimum
I	inspired
O	outside
o	reference
P, p	constant pressure
P	plasma
r	radiative
s	skin
T	tidal volume
V	constant volume
V	venous
tot	total
R	reaction
v	vaporization
w	wall

Biomedical Applications of Heat and Mass Transfer

chapter 1

Energy and Its Transformation

INTRODUCTION

In this introductory chapter we will discuss the relationships between different forms of *energy* and will present the major principles of equilibrium thermodynamics. While the application of these principles is limited in situations where the transfer of material and energy is occurring, it is desirable that the student comprehend these processes. It is necessary to have some working knowledge of the relations between energy, heat, and work in order to perform the basic material and energy balance computations that are the necessary foundation to the study and consideration of the rate processes of heat and mass transfer. Also, the characterization of systems in which chemical reactions are occurring requires the presentation of some elements of chemical thermodynamics and equilibrium.

We begin with a summary of the fundamental physical quantities, and their units, dimensions, and interrelations. We then proceed directly to consideration of the different forms of energy.

FUNDAMENTAL PHYSICAL QUANTITIES

The fundamental physical quantities needed for a discussion of energy and its transformation are summarized in Table 1.1. The basic quantities are presented in the three most commonly used unit systems, and the basic dimensions are presented in both the FLt and MLt (force-length-time and mass-length-time) systems. While one set of units and dimensions might be more convenient or aesthetically pleasing, a mastery of all the combinations of Table 1.1 is desirable to follow work being done in the interdisciplinary fields of engineering and life sciences.

Some discussion of the relationship between *force* and *mass* is necessary, since this continues to be a confusing point for students.

Table 1.1. Fundamental Physical Quantities

Physical Quantity	Units			Dimensions	
	English	*cgs*	*mks*	*Mass*	*Force*
Mass	lb_m	gm	kgm	M	Ft^2/L
Length	ft	cm	m	L	L
Time	sec	sec	sec	t	t
Force	lb_f	dyne	newton	ML/t^2	F
Work	ft-lb_f	erg	joule	ML^2/t^2	FL
Energy	BTU	cal	kcal	ML^2/t^2	FL
Power	ft-lb_f/sec	watt	kwatt	ML^2/t^3	FL/t
Temperature	°F	°C	°C	T	T
Heat	BTU	cal	kcal	ML^2/t^2	FL

The basic relation is given by Newton's law, which says that the amount of force required to give a certain amount of mass a given acceleration is proportional to the product of the mass and the acceleration, that is,

$$F \propto m\,a \tag{1.1}$$

where

F = force (measured in lb_f, dynes, or newtons)
m = mass (measured in lb_m, gm, kgm)
a = acceleration (measured in ft/sec^2, cm/sec^2, m/sec^2)

Another way of regarding this relationship is to define the force required to hold a certain mass stationary in the presence of a gravitational field—in other words, the definition of *weight*. Since it is convenient to have a numerical equivalence between force and mass on the earth's surface, the conversion factor—that is, the proportionality constant in (1.1)—can be so chosen, and its numerical value becomes equal to the acceleration of the earth's gravity field at sea level. Then,

$$W = m\,g/g_c \tag{1.2}$$

where

W = weight
m = mass
g = acceleration provided by the gravitational field (g_0 = 32.174 ft/sec^2 at earth's sea level)
g_c = 32.174 lb_m ft/lb_f sec^2

and

$$F = m\,a\,/\,g_c \tag{1.3}$$

Table 1.2 presents several numerical conversion factors to interrelate the various units and measures employed in this text.

Table 1.2. Basic Conversions

1 lb_f = 32.174 lb_m ft/sec^2	1 dyne = 1 gm cm/sec^2
1 BTU = 778 ft-lb_f	1 newton = 1 kgm m/sec^2
1 HP = 550 ft-lb_f/sec	1 joule = 1 newton m
1 lb_m = 454 gm	1 erg = 1 dyne cm
1 ft = 30.48 cm	1 cal = 4.184 joules
1 BTU = 252 cal = 0.252 kcal	1 watt = 1 joule/sec

FORMS OF MECHANICAL AND THERMAL ENERGY

The physical result of a force F acting through a distance L may be defined as *work*. This is the simplest form of energy in transition, in that when work is exerted on an object or on a system, its energy content is changed. The simplest physical example is the lifting of a weight with the resultant increase in *potential energy*, and this in turn provides our first definition of potential energy, which will be given the symbol ϕ.

> ϕ = potential energy = the work which can be extracted or is required when a system or object is moved in a potential field—for example, a mass moved in a gravitational field or a point charge moved in an electrical field, relative to some point of reference.

Potential energy can be measured in ft-lb_f, calorics, BTU's, ergs, or in any combination of FL units. It may also be thought of as energy content resulting from *position*.

Kinetic energy may be defined as energy resulting from *relative motion*. Newton's law may be used here to show that it is necessary to do work to change the kinetic energy of a given amount of mass.

Since for constant mass,

$$g_c F = m\,a = m\,dv/dt \tag{1.4}$$

Multiplying both sides by a differential distance dx

$$g_c F dx = m\,dv/dt\,dx = m\,v\,dv \tag{1.5}$$

and integrating, the definition of work results, and further,

$$\text{Work} = \int_1^2 F\,dx = \frac{m\,v_2^2}{2g_c} - \frac{m\,v_1^2}{2g_c} = \Delta K \tag{1.6}$$

where K = kinetic energy.

Kinetic energy may then be expressed and measured in the same units as potential energy, as an examination of Table 1.1 will show.

In addition to these two mechanical forms of energy, matter can possess relative energy as a consequence of its *molecular state*. One measure of relative molecular activity is of course *temperature*. Another parameter which will differentiate molecular state in a macroscopic sense is *composition*. This type of molecular state energy is given the label *internal energy* and given the symbol U. It is appropriate here to introduce the symbol $\hat{}$, which will be used to denote per unit mass so that $\hat{\phi}$, $\hat{K}$, and $\hat{U}$ will stand respectively for potential, kinetic, and internal energy per unit mass, or in other words for specific energy content.

We have seen that the energy content of a system may be changed by doing work on the system. Obviously one form of energy can be changed to another—perhaps the simplest transformation being the case of the conversion of potential energy to kinetic energy associated with falling bodies. At this point it is necessary to carefully define the term *system* as that portion of all matter that is under consideration. A *closed system* will be defined as a system across the boundaries of which no material crosses. An *open system* will be defined as a system in which material is allowed to cross its boundaries. An *isolated system* is a closed system with no ability to perform work or transfer energy.

Another simple way in which the energy content of a system may be changed is by the phenomenon of *heat*, which is defined as the transfer of energy resulting from a difference in *temperature*. The term *heat transfer* then is seen to be redundant, but it is a commonly used and accepted phrase in modern engineering technology and will be used somewhat remorsefully in this text.

The units and dimensions of all of these basic energy quantities are summarized in Tables 1.1 and 1.2.

FIRST LAW OF THERMODYNAMICS AND CONSERVATION OF ENERGY

The *first law of thermodynamics* may be regarded either as a statement of the conservation of energy or as a definition of internal energy as a function purely of the thermodynamic state of the system. For a single phase, one-component, closed system fixed in space the first law may be stated in its simplest form as

$$\Delta\hat{U} = \hat{Q} - \hat{W} \tag{1.7}$$

or in differential form

$$dU/dt = \dot{Q} - \dot{W} \tag{1.8}$$

where

$$U = m_{\text{tot}}\,\hat{U} \tag{1.9}$$

$$\dot{Q} = \frac{d}{dt}(m_{\text{tot}}\,\hat{Q}) \tag{1.10}$$

$$\dot{W} = \frac{d}{dt}(m_{\text{tot}}\,\hat{W}) \tag{1.11}$$

where

m_{tot} = the total amount of mass in the system
$\hat{Q}$ = amount of energy transferred as heat into the system per unit mass of material in the system
$\hat{W}$ = amount of energy transferred out of the system (by virtue of it doing external work) per unit mass

An important consequence of the first law is that the internal energy change resulting from some process will be independent of the thermodynamic path followed by the system, and of the paths followed by the processes described on the right side of (1.7). In turn, the rate at which the internal energy content of the system changes is dependent only on the rates at which heat is added and work is done.

For a system containing more than one chemical specie, the first law must be modified to include the effect of molecular interaction. Redefinition of the total internal energy as defined by (1.9) accomplishes this.

$$U = \sum_i (m_i\,\hat{U}_i + \Delta U_{ij}) \tag{1.12}$$

where

$\hat{U}_i$ = internal energy of species i per unit mass of i
m_i = mass of species i
ΔU_{ij} = internal energy change on mixing

If we allow the closed system to be moved in a potential field, it is necessary to include terms to account for changes in the potential and kinetic energy of the system. Equation 1.8 then becomes

$$\frac{d}{dt}(\phi + K + U) = \dot{Q} - \dot{W} \tag{1.13}$$

where

$$\phi = \hat{\phi}\, m_{\text{tot}}$$
$$K = \hat{K}\, m_{\text{tot}}$$

Equation 1.13 demonstrates how the various energy contents of a closed system may be interchanged (left side) and how the transfer

of energy from the system to the surroundings (right side) can affect the system.

Example 1.1. Falling Body

If friction is neglected, (1.13) may be directly applied to this case, with the result that $\dot{Q} = 0$, $\dot{W} = 0$, $dU/dt = 0$, and

$$d\phi/dt = -dK/dt \tag{1.14}$$

Example 1.2. Braking Automobile

The rate at which an automobile may be decelerated will be related to the rate at which heat is dissipated through the braking system. At level ground and neglecting any gross temperature changes, (1.13) becomes

$$dK/dt = \dot{Q} \tag{1.15}$$

It should be noted that both quantities in (1.15) will be negative.

Example 1.3. Space Capsule Reentry

In this situation every term in (1.13) is significant. The capsule decelerates, gains temperature, loses potential energy, transfers heat, and is appreciably worked on through the frictional forces resulting from the viscous atmosphere. In other words, it loses kinetic and potential energy, and gains internal energy while experiencing negative $\dot{Q}$ and $\dot{W}$.

Example 1.4. Resting Humans

If the small mass interchange with the surroundings is neglected, an approximate description of a resting human body can be written from (1.13). (We shall treat this situation more exactly in later chapters.) Since there are no potential or kinetic energy changes and no work is being done external to the system, (1.13) becomes

$$dU/dt = \dot{Q} \tag{1.16}$$

where both terms are negative. The left-hand term represents the conversion of stored chemical energy (basal metabolism), and the right side represents the amount of energy per unit time being transferred to the environment (about 70 kcal/hr for an average-sized adult male). If there are no internal temperature changes, this equation can be used as the basis for measuring the basal metabolic energy conversion rate for human subjects.

For many thermodynamic processes in closed systems the only significant energy changes are internal energy changes, and the only significant work done by the system in the absence of friction is the work of pressure-volume expansion. For this special case, (1.7) may be rewritten as

$$d\hat{U} = \delta\hat{Q} - P\,d\hat{V} \tag{1.17}$$

where $\delta\hat{Q}$ represents a discrete (not differential) amount of heat and $\hat{V}$ represents the specific volume (volume/unit mass) of the system. Because this simple system or situation arises (at least approximately) so frequently, it becomes convenient to introduce a new function $\hat{H}$, the *enthalpy*, defined as

$$\hat{H} = \hat{U} + P\,\hat{V} \tag{1.18}$$

and to rewrite (1.17) as

$$d\hat{H} = \delta\hat{Q} + \hat{V}\,dP \tag{1.19}$$

so that for processes and systems with constant pressure (isobaric) that also follow the conditions of (1.17)

$$d\hat{H} = \delta\hat{Q} \tag{1.20}$$

so that the enthalpy, a function only of the thermodynamic state as defined by (1.18), becomes related only to the heat evolved or absorbed by the system. Next, one may define the overall enthalpy content of the system as a function of the enthalpy content of the contained species and write

$$H = m_{\text{tot}}\,\hat{H} = \Sigma(m_i\,\hat{H}_i + \Delta H_{ij}) \tag{1.21}$$

and the related molar form

$$H = \Sigma(n_i\,\widetilde{H}_i + \Delta H_{ij}) \tag{1.22}$$

where n_i = the number of moles of species i, and the symbol ~ denotes the specific molar quantity, that is, per unit mole.

For systems with constant volume, the enthalpy change will be equal to the internal energy change, as stated by (1.18).

Since the enthalpy is a state function, it is necessary to measure it relative to some reference state. The usual practice is to choose the pure element in its natural molecular form at a temperature of 298° K and a pressure of one atmosphere (atm) as the standard state for each component and to designate the enthalpy at this state as zero for pure elements. For chemical compounds the enthalpy change associated with the formation of one mole of the compound in its standard state from its constituent elements in their standard states is used as the reference value. This reference value for compounds is called the *standard enthalpy of formation* (or more loosely, the heat of formation), since it corresponds to the energy (or heat) that would be required in forming the compound from its elements. Most enthalpies of formation are negative, since energy is often given off during formation. A positive value implies that the compound will be formed in its reference state only by adding energy to the system. Typical values of the standard enthalpy of formation, denoted hereafter by the symbol $\Delta\widetilde{H}_f^\circ$ are given in Table 1.3.

Table 1.3. Standard Enthalpy of Formation, $\Delta\widetilde{H}_f^o$ (kcal/mole), $T = 298°K, P = 1$ atm.

Gases		*Liquids*	
H_2O	-57.79	CH_3OH	-57.02
SO_2	-70.96	C_6H_6	11.72
NO	21.60	CCl_4	-33.30
CO_2	-94.05	H_2O	-68.32
CO	-26.41		
CH_4	-17.98	*Solids*	
C_2H_6	-20.24		
O_2	0.0	$Ca(OH)_2$	-235.6
O	59.1	Al_2O_3	-399.1

These values of $\Delta\widetilde{H}_f^o$ may be used to calculate the enthalpy changes that result from chemical reaction when the products and reactants are both brought to 298°K (25°C) and 1 atm. For example, if CO is burned to produce CO_2 according to the following,

$$CO + 1/2\ O_2 \longrightarrow CO_2 \tag{1.23}$$

then the standard enthalpy change, or heat required at constant temperature and pressure or the *heat of reaction*, may be computed as

$$\Delta\widetilde{H}_R^o\,(\text{reaction}) = \Sigma\widetilde{H}_f^o\,(\text{products}) - \Sigma\widetilde{H}_f^o\,(\text{reactants}) \tag{1.24}$$

From Table 1.3,

$$\Delta\widetilde{H}_R^o\,(\text{reaction}) = (-94.05) - 0 - (-26.41)$$

$$\Delta H_R^o = -67.64 \text{ kcal/mole } CO_2 \text{ produced}$$

That is, if one gram mole (28 gm) of CO is completely burned, 67,640 cal of energy will be evolved as heat, provided that the entire reaction were to take place at 25°C. For the special category of combustion with oxygen, the standard heat of reaction is called the *heat of combustion.*

In most biological systems, chemical reactions occur essentially at isothermal conditions, and the standard enthalpy change of reaction is an important quantity. Table 1.4 lists some important examples

Table 1.4. Selected Values of $\Delta\widetilde{H}_R^o$ (kcal/mole)

Reactants	*Products*	$\Delta\widetilde{H}_R^o$
Glucose, O_2	CO_2, H_2O	-673.0
Sucrose, O_2	glucose, fructose	-4.8
Glucose 6-phosphate, H_2O	glucose, H_3PO_4	-3.0
Lactic acid, O_2	CO_2, H_2O	-326.0
Palmitic acid, O_2	CO_2, H_2O	-2,380.0
NaOH, HCl	NaCl, H_2O	-13.8

and demonstrates that heats of reaction varying through several orders of magnitude are common in biological systems. It should be emphasized, however, that these values represent the energy released per mole of reactant only when the reaction has proceeded to completion.

At this juncture it becomes necessary to demonstrate how the variation of energy content (or enthalpy) with temperature can be accounted for. The thermodynamic function, *heat capacity*—denoted with the symbol $\hat{C}$ and having the units of energy/mass temperature—is used and will be defined for both constant pressure and constant volume processes as follows, with the subscripts p and v accordingly:

$$\hat{C}_p = (\partial \hat{H}/\partial T)_P$$
$$\hat{C}_v = (\partial \hat{U}/\partial T)_v \qquad (1.25)$$

with the molar values defined similarly as

$$\widetilde{C}_p = (\partial \widetilde{H}/\partial T)_p$$
$$\widetilde{C}_v = (\partial \widetilde{U}/\partial T)_v \qquad (1.26)$$

The heat capacity of liquid water at 4°C on a mass basis has a value of 1.0 kcal/kgm °C, (1.0 BTU/lb_m °F) and is often used as a reference quantity to define the specific heat of liquids by expressing the heat capacity of the substance in question as a ratio of that quantity to the heat capacity of water.

Given the heat capacity of water, the enthalpy change that results (or the amount of heat needed) when the temperature of a given mass of water is changed isobarically may be computed by combining (1.25) and (1.20) as

$$\dot{Q} = \Delta H = m_{tot} \, \Delta \hat{H} = m_{tot} \int_{T_1}^{T_2} \hat{C}_p \, dT \qquad (1.27)$$

or

$$\dot{Q} = m_{tot} \, (1.0) \, (\Delta T) \text{ kcal} \qquad (1.28)$$

since the heat capacity of water is relatively constant over small temperature ranges.

Example 1.5. Body Temperature Regulation

The average heat capacity of the human body is approximately 0.86 kcal/kgm °C. This implies that the capacity of the body to store thermal energy is somewhat less than that of water. For example, if the temperature of the body is lowered by 1°C, the energy required to completely reheat the body to its original temperature is computed by

$$\hat{Q} = \Delta \hat{H} = (0.86) \, (1.0) \text{ kcal/kgm} \qquad (1.29)$$

For a 70-kgm body,

$$\dot{Q} = m_{tot}\,\hat{Q} = 70 \text{ kgm } (0.86 \text{ kcal/kgm}) = 60 \text{ kcal} \qquad (1.30)$$

The amount of pure glucose that would have to be completely oxidized to provide this amount of energy can be easily computed from the preceding considerations.

$$m_{tot} = \dot{Q}\widetilde{M}/\Delta\widetilde{H}_R = \frac{60 \text{ kcal } (180 \text{ gm/mole})}{673 \text{ kcal/mole}} = 16 \text{ gm} \qquad (1.31)$$

where $\widetilde{M}$ = molecular weight.

Example 1.6. Calorific Oxygen Equivalents

One way to measure metabolic conversion rates in the body is to measure the rate of oxygen consumption. This principle may be illustrated by computing the equivalent calorific oxygen content for glucose combustion.

From Table 1.4 it is seen that the enthalpy change of reaction for glucose is -673 kcal/mole. From the balanced stoichiometric reaction

$$C_6H_{12}O_6 \text{ (glucose)} + 6\,O_2 \longrightarrow 6\,H_2O + 6\,CO_2$$

we see that 6 moles of oxygen are required to completely burn each mole (180 gm) of glucose. From this data we can compute the following important values:

Liters of O_2/gm fuel = 6 × 22.4/180 = 0.75 (at 0° C)
Liters of CO_2 produced/gm fuel = 0.75
Respiratory quotient = 1.0 liters of CO_2/liter of O_2
Kcal produced/gm fuel = 673/180 = 3.75
Kcal equivalent of 1 liter of O_2 = 3.75/0.75 = 5.00

Although this calculation is performed here only for glucose, similar computations may be performed easily for any foodstuff. When this is done and averaged values are computed, Table 1.5 results.

Table 1.5. Energy Relationships

Quantity	*Carbohydrate*	*Fat*	*Protein*
Liters of O_2 used/gm*	0.75	2.03	0.97
Liters of CO_2 produced/gm	0.75	1.43	0.78
Respiratory quotient	1.00	0.71	0.80
Kcal produced/gm	4.10	9.30	4.10
Calorific equivalent of 1 liter O_2	5.47	4.60	4.23

*At 0° C, 1 atm.

For an average mixed diet a representative value for the calorific oxygen equivalent is 4.825 kcal/liter of oxygen consumed. In the computations involved in Table 1.5 it is necessary to correct the heats of reaction involved due to the fact that in vivo the reactions do not go completely to the stoichiometric products. For example, in a bomb calorimeter, the kcal produced per gm of protein is 5.3, whereas in the human body the value is close to 4.1, most likely indicating that complete oxidation does not occur.

Example 1.7. Energy Expended in Respiratory Gases

If dry air at 20° C is inhaled and the body in turn exhales a water-saturated gas mixture at 37° C, some energy will be lost. The rate of energy loss may be approximated by comparing the enthalpy of the two streams. If we assign an average heat capacity of 0.25 kcal/kgm °C for the gases and take a flow rate of 6,000 ml/min, the following analysis ensues:

$$\dot{Q} = \Delta H = H_{\text{out}} - H_{\text{in}} = \Delta H_{\text{gas}} + \Delta H_{\text{water}}$$

$$\dot{Q} = \text{energy used to heat the dry gas} + \text{energy used to vaporize the water}$$

$$\dot{Q} = w_g \hat{C}_p (T_{\text{in}} - T_{\text{out}}) + w_L (-\Delta \hat{H}_v) \tag{1.32}$$

where

$\dot{Q}$ = rate at which energy is lost
w_g = mass flow rate of the dry gas
w_L = mass rate at which water is vaporized
$-\Delta \hat{H}_v$ = heat of vaporization of water at 37°C = 0.54 kcal/gm

Employing the ideal gas law, w_g can be computed.

$$w_g = \frac{6{,}000 \text{ ml/min } (273^\circ \text{C}) (28.9 \text{ gm/mole})}{22{,}400 \text{ ml/mole } (293^\circ \text{C})} = 7.2 \text{ gm/min} \tag{1.33}$$

The amount of water vaporized may be computed if the partial pressure of water at saturation at 37° C (47 mm Hg) is known.

$$w_L = \left(\frac{6{,}000 \text{ ml dry gas/min}}{22{,}400 \text{ ml/mole}}\right)\left(\frac{47 \text{ ml water}}{760\text{-}47 \text{ ml dry gas}}\right)\left(\frac{273}{293}\right) \times 18 \text{ gm/mole} = 0.27 \text{ gm/min} \tag{1.34}$$

Substituting these values into (1.33),

$$\dot{Q} = (7.2 \text{ gm/min}) (0.24 \text{ cal/gm }^\circ\text{C}) (-12^\circ \text{C}) + (0.27 \text{ gm/min}) (540 \text{ cal/gm}) = -21 - 140 = -161 \text{ cal/min} = -0.161 \text{ kcal/min} \tag{1.35}$$

or 9.7 kcal/hr or 230 kcal/day.

This example demonstrates that an appreciable rate of energy loss can be experienced simply by normal breathing.

SECOND LAW OF THERMODYNAMICS AND THE CONCEPT OF REVERSIBILITY

The first law of thermodynamics defines internal energy as a state function and provides a formal statement of the conservation of energy. However, it provides no information about the direction in which processes can spontaneously occur, that is, the reversibility aspects of thermodynamic processes. It cannot, for example, tell how cells can perform work while existing in an isothermal environment. It gives no information about the inability of any thermodynamic process to convert heat into mechanical work with full efficiency, or any insight into why mixtures cannot spontaneously separate or unmix themselves. An experimentally derived principle to characterize the availability of energy is required to do this. This is precisely the role of the *second law of thermodynamics.*

The second law defines the fundamental physical quantity *entropy* as a randomized energy state unavailable for direct conversion to work. It also states that all spontaneous processes, both physical and chemical, proceed to maximize entropy, that is, to become more randomized and to convert energy into a less available form. A direct consequence of fundamental importance is the implication that at thermodynamic equilibrium the entropy of a system is at a relative maximum; that is, no further increase in disorder is possible without changing by some external means (such as adding heat) the thermodynamic state of the system. A basic corollary of the second law is the statement that the sum of the entropy changes of a system and that of its surroundings must always be positive, that is, the universe (the sum of all systems and surroundings) is constrained to become forever more disordered and to proceed toward thermodynamic equilibrium with some absolute maximum value of entropy. From a biological standpoint this is certainly a reasonable concept, since unless gradients in concentration and temperature are forcibly maintained by the consumption of energy, organisms proceed spontaneously toward the biological equivalent of equilibrium—death.

A formal statement of the second law may be written as

$$\Delta \hat{S}_{\text{system}} = \hat{Q}/T + \hat{R}/T \tag{1.36}$$

$$\Delta \hat{S}_{\text{surroundings}} = -\hat{Q}/T \tag{1.37}$$

where

$\hat{S}$ = entropy/unit mass (energy/mass temperature)
$\hat{Q}$ = energy added to the system from the surroundings in the form of heat/unit mass
$\hat{R}$ = energy lost from the system by friction or other irrevers-

ibilities. This term is always positive. It is manifested either by a temperature increase, a velocity decrease, a pressure decrease, or some combination of these. It represents a loss in the system's capability to perform mechanical work and therefore is always responsible for an increase in the entropy of the system.

The sum of these two statements, and the corollary implied, is

$$\Delta \hat{S}_{\text{system + surroundings}} = \hat{R}/T \geqslant 0 \tag{1.38}$$

which is the mathematical statement of the results described above.

For closed systems (all we have considered so far) the first and second laws may be combined, independent of irreversibilities, to provide some useful relationships between thermodynamic quantities. Recalling the differential form of the first law (1.7),

$$d\hat{U} = \delta \hat{Q} - \delta \hat{W} \tag{1.39}$$

and using a restatement of (1.36)—without the subscript, which will be implied—the following results.

$$\delta \hat{Q} = T\, d\hat{S} - \delta \hat{R} \tag{1.40}$$

Further, observing that the mechanical work obtainable from a closed system in the absence of electrical or magnetic effects may be expressed as

$$\delta \hat{W} = P\, d\hat{V} - \delta \hat{R} \tag{1.41}$$

that is, the mechanical work available will be that from pressure-volume expansion minus that lost in friction. One then obtains

$$d\hat{U} = T\, d\hat{S} - P\, d\hat{V} \tag{1.42}$$

and subsequently, recalling the definition of enthalpy in (1.18),

$$d\hat{H} = T\, d\hat{S} + \hat{V}\, dP \tag{1.43}$$

These relations can be used to show that for ideal gases ($P\widetilde{V} = RT$)

$$\hat{C}_p - \hat{C}_v = R \text{ (the gas constant)} \tag{1.44}$$

and also that the internal energy is independent of the pressure, that is,

$$(\partial \hat{U}/\partial P)_T = 0 \tag{1.45}$$

both of which are good exercises for the reader and important thermodynamic relations.

Isentropic processes are those for which there is no change in entropy, that is, $\Delta \hat{S} = 0$. This may occur if both $\hat{Q} = 0$ (*adiabatic* process), and $\hat{R} = 0$ (*frictionless* or *reversible* process), or if $\hat{Q} = -\hat{R}$; that is, the energy lost by friction is exactly balanced by a heat output. This case often occurs, at least to a good degree of approximation in such "fast" processes as expansions through nozzles.

A summary of some of these terms for closed systems follows:

Adiabatic	$\hat{Q} = 0$	no heat transfer
Reversible	$\hat{R} = 0$	no frictional losses
Isentropic	$\Delta S = 0$	constant entropy
Isothermal	$dT = 0$	constant temperature
Isobaric	$dP = 0$	constant pressure
Isochoric	$dV = 0$	constant volume

The applicability of the second law can be illustrated with some simple examples.

Example 1.8. Mixing of Two Gas Samples

Suppose two samples of an ideal gas are initially separated within a volume V_0 into two fractions aV_0 and $(1 - a)V_0$, with $a < 1$. Let the pressure and temperature in each compartment be held at P_0 and T_0. We want to compute the entropy change when the samples mix spontaneously and to show that the reverse process cannot occur.

If no heat or external work is transferred to the mixing chambers, the first law tells us that the internal energy content will be constant ($d\hat{U} = 0$); then from (1.42) we can obtain the relation between entropy change and specific volume change for each portion of the gas.

$$(\partial \hat{S}/\partial \hat{V})_T = P/T \tag{1.46}$$

That is, the rate of entropy change with volume is given for this situation by the ratio of pressure and temperature. For an ideal gas,

$$P V = n R T \tag{1.47}$$

or

$$P/T = n R/V \tag{1.48}$$

Substituting these into (1.46), and integrating,

$$\Delta S = \int_{V_1}^{V_2} (\partial S/\partial \hat{V}) dV = \int_{V_1}^{V_2} (n R/V) dV = n R \ln (V_2/V_1) \tag{1.49}$$

The entropy change of the a fraction will then be

$$\Delta S_a = a n R \ln (V_0/aV_0) \tag{1.50}$$

and that of the $(1 - a)$ fraction will be

$$\Delta S_{(1-a)} = (1 - a) n R \ln [V_0/(1 - a) V_0] \tag{1.51}$$

and the sum of these, or the entropy change of the entire system, will be

$$\Delta S = \Delta S_a + \Delta S_{(1-a)} = n R \{a \ln (1/a) + (1 - a) \ln [1/(1 - a)]\} \tag{1.52}$$

Since $a < 1$, ΔS will always be positive for such a mixing process. According to the corollary of the second law, while this process can occur spontaneously, the reverse cannot.

Example 1.9. Heat Exchange between Two Blocks

Suppose that two blocks of the same material and equal mass of different temperatures T_h and T_c are brought into contact and insulated from the surroundings. It is desired to determine the relation between the resulting temperature and entropy changes and to draw some conclusion about the spontaneity of the process.

Once again the internal energy of the system (the two blocks) will remain constant, and (1.25) and (1.43) can be combined to yield

$$(\partial \hat{S}/\partial T)_P = \hat{C}_p/T \tag{1.53}$$

so that, for each block

$$\Delta \hat{S} = \int_{T_1}^{T_2} (\hat{C}_p/T)\, dT = \hat{C}_p \ln (T_2/T_1) \tag{1.54}$$

For the hot block

$$\Delta \hat{S}_h = \hat{C}_p \ln (T/T_h) \tag{1.55}$$

and for the cold

$$\Delta \hat{S}_c = \hat{C}_p \ln (T/T_c) \tag{1.56}$$

where T is the final mutual temperature of the two blocks after a long time. T can be computed easily if the two blocks have the same heat capacities and equal mass as

$$T = (T_h + T_c)/2 \tag{1.57}$$

The total entropy change of the system will then be

$$\Delta \hat{S} = \Delta \hat{S}_h + \Delta \hat{S}_c = \hat{C}_p \ln [T^2/(T_h T_c)] \tag{1.58}$$

Substituting (1.57), we observe that

$$\ln \left(\frac{T_h^2 + 2T_h T_c + T_c^2}{4T_h T_c} \right) \geqslant 0 \text{ if } (T_h - T_c) \geqslant 0 \tag{1.59}$$

That is, $\Delta \hat{S}$ will be positive and the process can occur spontaneously only if T_h is greater than T_c. Since negative overall entropy changes are prohibited by the second law, the reverse process cannot occur. This is another way of stating or demonstrating that heat cannot flow from a cooler to a warmer temperature.

The computation of absolute entropies for use in thermodynamic calculations is based on the *third law of thermodynamics*, which states that the entropy of a pure substance as a perfectly ordered crystal at a temperature of absolute zero (0° K) is zero ($\hat{S} = 0$). Using this as a reference point, and employing relations similar to (1.53) and (1.46), absolute entropies of solids, liquids, and gases may be computed as functions of temperature and pressure. An obvious consequence of the third law is that gases will have higher entropy in general than liquids, and liquids will have higher entropy than solids. Since gases are more disordered than either, this is an expected result. Since, on the other hand, a crystal at absolute zero is perfectly ordered, we are not surprised to find that it has the lowest possible value of entropy.

FREE ENERGY AND THE CONCEPT OF EQUILIBRIUM

Since most biological and physiological systems operate at nearly constant temperatures, a most useful thermodynamic quantity would be a measure of that portion of the total energy of a system available to perform work under isothermal conditions—its *free energy*. In one sense, this type of energy would be the inverse of entropy, since entropy is a measure of unavailable or degraded energy. Such a thermodynamic quantity was exactly defined by Gibbs and given the symbol $\hat{G}$,

$$\hat{G} = \hat{H} - T\hat{S} \tag{1.60}$$

Recalling the definition of H from (1.18),

$$\hat{G} = \hat{U} + P\hat{V} - T\hat{S} \tag{1.61}$$

we see that $\hat{G}$ indeed represents that component of the system's total thermal energy and flow energy remaining after the unavailable energy (entropy) is subtracted.

Employing (1.43) for a closed one-component system, we write

$$d\hat{G} = \hat{V}\,dP - \hat{S}\,dT \tag{1.62}$$

At constant pressure and temperature this may be arranged, using the first and second laws, as

$$d\hat{G} = \delta\hat{Q} - T\,d\hat{S} = -\hat{R} \tag{1.63}$$

and for a reversible, isothermal, isobaric, closed system

$$d\hat{G} = 0 \tag{1.64}$$

These two relationships—(1.63) and (1.64)—lead to the following important conclusions:

1. At constant T and P, $\hat{G}$ can decrease or remain constant, but never increase, since $\hat{R}$ is always positive or zero.
2. When a system reaches equilibrium ($d\hat{S} = d\hat{G} = dT = dP = 0$) the free energy $\hat{G}$ is at a minimum.

3. Since T is always positive, the entropy S will be at a maximum at equilibrium.

If we consider a process operating between two thermodynamic states,

$$\Delta \hat{G} = \Delta \hat{H} - T \Delta \hat{S} \tag{1.65}$$

we observe that the more negative $\Delta \hat{H}$ is (the more heat being lost to the surroundings), the more negative $\Delta \hat{G}$ is. Also, the more positive $\Delta \hat{S}$ is, the more negative $\Delta \hat{G}$ is, and in turn the more spontaneous is the process. We see that the free energy combines the tendency toward minimum entropy and toward minimum enthalpy to predict the energy changes and the degree of spontaneity of a system.

Example 1.10. Vaporization of Water

If one mole of liquid water is totally vaporized, $\Delta \widetilde{H}_v$ = 9.59 kcal/mole, and $\Delta \widetilde{S}$ = 25.7 cal/mole°K. From (1.65) ΔG will be negative as long as $T \Delta \widetilde{S} > \Delta \widetilde{H}$, or if $T >$ 373°K (100°C). At temperatures above 100°C this process will proceed spontaneously toward equilibrium. At T = 100°C, $\Delta \widetilde{G}$ = 0 and the process is at equilibrium (and is also reversible).

Earlier in the discussion of enthalpies of formation the concept of a standard state, at which the enthalpy of pure compounds was set equal to zero, was discussed. This device is equally necessary and useful for free energy computations. We again refer to the *free energy of formation* of a compound and denote it by the symbol $\Delta \widetilde{G}_f^o$. This quantity refers to the amount of free energy required to form the compound in question from its constituent elements in their respective standard states. Some sample values are given in Table 1.6.

Table 1.6. Selected Values of Free Energy of Formation, $\Delta \widetilde{G}_f^o$ (kcal/mole), T = 298°K, P = 1 atm

Gases		*Liquids*	
H_2O	-54.64	CH_3OH	-39.73
SO_2	-71.79	C_6H_6	29.76
NO	20.72	CCl_4	-16.4
CO_2	-94.26		
CO	-32.81	*Solids*	
CH_4	-12.14	$Ca(OH)_2$	-214.3
O_2	0.0	Al_2O_3	-376.8

Once again we may consider the energy changes that occur during chemical reaction. If all the products and reactants are brought to their respective standard states, we may define a *standard free energy change*

$$\Delta \widetilde{G}^o = \Sigma \Delta \widetilde{G}_f^o \text{ (products)} - \Sigma \Delta \widetilde{G}_f^o \text{ (reactants)} \tag{1.66}$$

Some standard free energy changes of interest in biological systems are presented in Table 1.7. If the standard free energy change is negative, this implies that the reaction will proceed spontaneously without the addition of energy. However, it should be stressed that the *rate* at which the reaction will proceed is independent of the free energy considerations and in fact is not predictable from the concepts of equilibrium thermodynamics. These concepts may be used to predict extent or direction, but not rate. Such factors as the presence or absence of catalysts, the molecular mechanisms of the intermediate chemical reactions, and the activation energies required to initiate the reaction must be taken into account to predict rates of reaction. The reader might wish to consult texts on chemical reaction kinetics, examples of which are listed at the end of this chapter.

Table 1.7. Selected Values of Free Energy Changes, $T = 298°K, P = 1$ atm

Oxidation (complete)	$\Delta\widetilde{G}^{\circ}$ *kcal/mole*
Glucose	-686.0
Lactic acid	-326.0
Palmitic acid	-2,338.0
Hydrolysis	
Sucrose + H_2O ⟶ glucose + fructose	-5.5
Rearrangement	
Glucose 1-phosphate ⟶ glucose 6-phosphate	-1.745

Equilibrium relationships for chemical reactions may now be discussed, using the concepts of free energy and the standard free energy change. First, it is necessary to standardize the strategy of computing concentrations or the relative amounts of each chemical species present. For species existing in gaseous form we define the *standard state* as the pure component gas in the ideal gas state at a temperature of 25°C and then define the *activity* of a gaseous species as the ratio of the partial pressure of the gas to a pressure of 1 atm. The partial pressure of the gaseous specie is defined as

$$p_i = y_i P \tag{1.67}$$

where

p_i = the partial pressure of component i
y_i = the mole fraction of component i
P = the total pressure in the system

Then, for gases

$$\overline{a_i} = p_i/P_o \tag{1.68}$$

where

$\overline{a_i}$ = the activity of component i
P_o = 1 atm

For species in the liquid state the standard state is chosen as a 1 molal water solution of the specie, at 25°C and 1 atm, and the activity is given by the ratio of the concentration of the specie to a 1 molal solution, or

$$\overline{a_i} = C_i/C_o \tag{1.69}$$

where C_i = the concentration of i in moles/liter.

Equipped with these definitions, the consideration of chemical reaction equilibria proceeds as follows. Consider an arbitrary chemical reaction,

$$a\,A + b\,B \longrightarrow c\,C + d\,D \tag{1.70}$$

where a, b, c, d denote the number of moles of each specie and A, B, C, D denote the species participating in the reaction.

We have seen how to compute the standard free energy change of the reaction from either enthalpy and entropy data or standard free energy of formation data—(1.65) or (1.66). If the chemical reaction takes place under isothermal conditions as in most biological situations; (1.62) may be used to write

$$d\widetilde{G} = \widetilde{V}\,dP \tag{1.71}$$

and for an ideal gas where $\widetilde{V} = RT/P$, upon integration,

$$\int_{\widetilde{G}^o}^{\widetilde{G}} d\widetilde{G} = RT \int_{P_o}^{P} dP/P \tag{1.72}$$

$$P_o = 1 \text{ atm} \tag{1.73}$$

gives

$$\widetilde{G} - \widetilde{G}^o = RT \ln (P/P_o) \tag{1.74}$$

For the chemical reaction of (1.70) this expression may be used to produce

$$\Delta\widetilde{G} = \Delta\widetilde{G}^o + RT \ln \left[\frac{(a_C)^c\,(a_D)^d}{(a_A)^a\,(a_B)^b}\right] \tag{1.75}$$

and if the standard pressure is 1 atm,

$$\Delta\widetilde{G} = \Delta\widetilde{G}^o + RT \ln \left[\frac{(p_C)^c\,(p_D)^d}{(p_A)^a\,(p_B)^b}\right] \tag{1.76}$$

Although the argument of the ln term might appear to have dimensions of pressure to some power for the case where $c + d \neq a + b$, it should be remembered that the term $(P_o)^{c+d-a-b}$, which is not written since $P_o = 1$, will supply the missing dimensions.

At equilibrium $\Delta \widetilde{G} = 0$,

$$\Delta \widetilde{G} = 0 \tag{1.77}$$

so

$$\Delta \widetilde{G}^o = -RT \ln \left[\frac{(p_C)^c \, (p_D)^d}{(p_A)^a \, (p_B)^b} \right] \tag{1.78}$$

and for liquids, replacing the activity with concentration ratios.

$$\Delta \widetilde{G}^o = -RT \ln \left[\frac{(C_C)^c \, (C_D)^d}{(C_A)^a \, (C_B)^b} \right] \tag{1.79}$$

with the same dimensional observation as in (1.76).

In both these cases it is convenient to define an *equilibrium constant* K as follows

$$\Delta \widetilde{G}^o = -RT \ln K \tag{1.80}$$

which relates the concentrations or partial pressures at equilibrium to the standard free energy change for the reaction at a given temperature T.

This relation shows that a very high conversion of reactants to products—that is, a large value of K—corresponds to a large negative free energy change and vice versa. A K value of 1.0 implies a standard free energy change of zero and essentially says that the stoichiometric relation is fulfilled.

Example 1.11. Glucose Rearrangement

An important example occurs as an intermediate step in the conversion and storage of glucose to glycogen. The reaction, glucose 1-phosphate ⟶ glucose 6-phosphate, when carried out at 25°C and a pH of 7.0, reaches an equilibrium in water solution such that the concentration of 1-phosphate is 0.001 M and 6-phosphate is 0.019 M. Using (1.80), $K = 19$ and the standard free energy change is -1.745 kcal/mole.

Example 1.12. Concentration of Stomach Acid

Glands in the digestive system produce approximately 2.5 liters/day of 0.16 N hydrochloric acid, using plasma concentration fluids as a raw material. The energy required to accomplish this concentration can be estimated, using (1.79). We shall consider the hydrogen ion concentration process, as adequate chloride is present in plasma to produce this level of acid concentration.

Concentration of H^+ in plasma: pH = 7.4, $[H^+] = 4 \times 10^{-8}$ equiv/liter
Concentration of H^+ in acid = 0.16 equiv/liter
Concentration ratio = $K = 4 \times 10^6$
$\Delta \widetilde{G} = RT \ln K = 1430 \ln (4 \times 10^6)$ cal/mole
= 9.5 kcal/mole acid
$\Delta \widetilde{G} = 9.5\ (0.16)$ kcal/liter of acid = 3.75 kcal/day

Considering that the efficiency of this process is probably less than 50%, these glands require about 8 kcal/day to operate—a substantial amount of energy.

Example 1.13. Electrochemical Processes and the Nernst Equation

The object in this example is to relate the free energy differences needed to sustain electrical potential gradients to predict the amount of work obtainable from electrochemical cells, and in addition to predict the amount of energy needed to maintain potential differences across such barriers as cell membranes. Also by combining some of these concepts, we shall see that voltage measurements can be used to determine concentration differences across interfaces.

First, we consider an electrochemical cell operating at constant pressure and temperature and rewrite (1.60) as

$$d\widetilde{G} = d\widetilde{H} - d(T\widetilde{S}) \tag{1.81}$$

using the definition of H and the first law,

$$d\widetilde{G} = -\delta \widetilde{W} + P\, d\widetilde{V} \tag{1.82}$$

Observing that the work that is obtainable in general is

$$\delta \widetilde{W} = P\, d\widetilde{V} + \delta \widetilde{W}_{\text{electrical}} \tag{1.83}$$

and that

$$\delta \widetilde{W}_{\text{electrical}} = n\, \mathfrak{F}\, \epsilon \tag{1.84}$$

where

n = number of moles of charge
$\mathfrak{F}$ = the charge on one mole of electrons, or 96,486 coulombs/mole
ϵ = potential difference, volts

For a standard cell, substituting into (1.83),

$$\Delta \widetilde{G}^o = -n\, \mathfrak{F} \Delta \epsilon^o \tag{1.85}$$

which is the relationship between the standard free energy difference and the standard voltage for an electrochemical cell.

Recalling (1.80) and combining this with (1.85) and (1.77),

$$\Delta \epsilon = \Delta \epsilon^o - (RT/n\mathfrak{F}) \ln K \tag{1.86}$$

at equilibrium, $\Delta \epsilon = 0$, and

$$\Delta \epsilon^o = RT/n\mathfrak{F} \ln K \tag{1.87}$$

Measuring $\Delta \epsilon^o$ in millivolts, T at 37°C, and substituting for R and $\mathfrak{F}$, the familiar Nernst equation results:

$$\Delta \epsilon^o = 61 \ln K \tag{1.88}$$

which enables one to predict the concentration gradients K by measuring voltage differences. In this regard, one could think of a voltmeter as a "free energy difference meter."

Example 1.14. Osmosis

The concept of free energy is instrumental in the understanding and analysis of the process of osmosis—defined as the passage of a solvent liquid through a semipermeable membrane, which is impermeable to certain of the solute molecules—and the associated establishment of an osmotic pressure across the membrane.

Suppose a solution of species A dissolved in a solvent of species B is enclosed in a tube having one end covered by a membrane permeable only by B, and this end is dipped into a solution of pure B, as shown in Figure 1.1. Fluid B will tend to diffuse through the membrane, driven by a concentration gradient, until a positive excess pressure develops inside the tube and brings the system to equilibrium. We begin the analysis of this situation by expressing the free energy of the solvent B inside the tube relative to its standard state as

$$\widetilde{G}_B = \widetilde{G}_B^o + RT \ln a_B \tag{1.89}$$

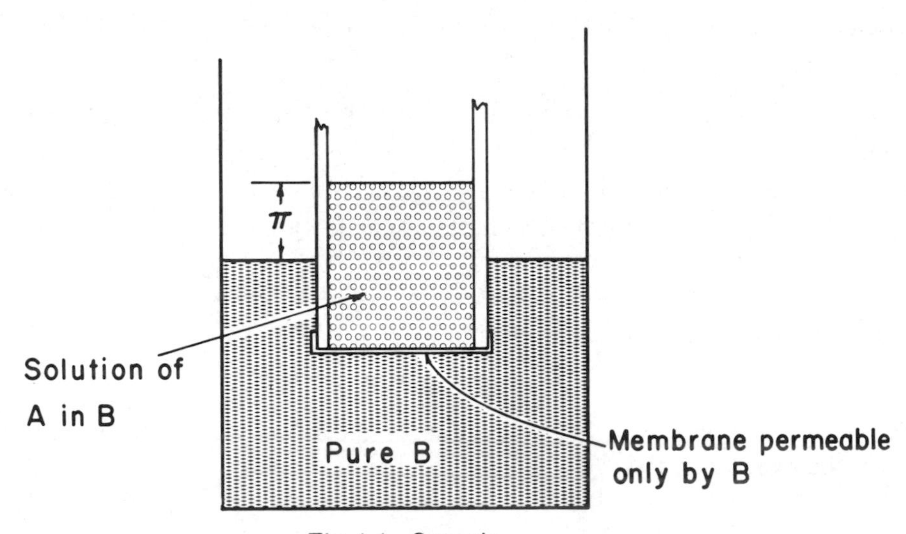

Fig. 1.1. Osmosis.

For convenience we will choose the standard state for this situation as pure B at the temperature of the system. Then,

$$\widetilde{G}_B = \widetilde{G}^o_B + RT \ln x_B \tag{1.90}$$

where

x_B = mole fraction of B in solution
$\widetilde{G}^o_B$ = free energy of pure B

With no other forces present, the flow of B through the membrane will continue until the concentration difference disappears. But, recalling (1.62) again, at constant temperature

$$d\widetilde{G} = V\, d\widetilde{P} \tag{1.91}$$

which implies that an increase in hydrostatic pressure (resulting from the extra B flowing into the tube) will increase the free energy of the contents of the tube by the amount

$$\Delta \widetilde{G} = \int_0^{\pi} \widetilde{V}\, dP = \pi \widetilde{V}_B \tag{1.92}$$

where

π = the hydrostatic head developed in the tube, depicted in Figure 1.1
$\widetilde{V}_B$ = the specific volume of B

Then, combining this with (1.90),

$$\widetilde{G}_B = \widetilde{G}^o_B + RT \ln x_B + \pi \widetilde{V}_B \tag{1.93}$$

and at equilibrium, where $\widetilde{G}_B = \widetilde{G}^o_B$,

$$\pi \widetilde{V}_B = -RT \ln x_B \tag{1.94}$$

If the mole fraction of the solute x_A is small,

$$\ln x_B = \ln (1 - x_A) \approx -x_A \tag{1.95}$$

Then, since

$$\widetilde{V}_B = V/n_B$$

and

$$x_A = n_A/n_B$$

gives

$$\pi = (RT/V)\, n_A \tag{1.96}$$

which says that the osmotic pressure that develops will, for small concentrations, be directly proportional to the number of osmotically active ions present in the solution. Equation 1.96 is, in fact, very similar to the ideal gas law.

This example demonstrates how a free energy difference may be interpreted as a driving force for the transfer of a given species, a concept that will be very important in following chapters.

SUMMARY

In this chapter the basic concepts necessary to describe the flow of energy in the biological world have been reviewed. Some feeling of the relative flows of energy on the earth should be held by the reader. Whereas the amount of energy transformed by man-made machines is on the order of 10^{16} kcal/yr, the flow of energy in the biological world is on the order of 10^{18} kcal/yr. Both of these quantities pale when compared with the solar energy flow received by the earth, which is on the order of 10^{21} kcal/year.

Bibliography

This chapter represents only a review of the basic principles of equilibrium chemical thermodynamics. The reader is urged to consult more complete texts for further details or for more complete derivation of many of the basic results employed in this chapter.

Two appropriate texts for life science students are:

Lehninger, A. L. *Bioenergetics*. Benjamin, 1965.

Mahan, B. H. *Elementary Chemical Thermodynamics*. Benjamin, 1964.

In addition the following texts on chemical kinetics are useful for engineering students:

Aris, R. *Chemical Reaction Analysis*. Prentice-Hall, 1965.

Smith, J. M. *Chemical Engineering Kinetics*. McGraw-Hill, 1956.

Also, for a more sophisticated treatment specifically developed for physiology students, the following is highly recommended:

Dowben, R. J. *General Physiology*. Harper and Row, 1969.

chapter 2

Material Balances in Living Systems

INTRODUCTION

The capability to account correctly for all the flows of material entering and leaving a physical, chemical, or biological system is essential to any further analysis of the system and is therefore of fundamental and primary importance. For example, the ability to "keep track" of the relative amounts of the various chemical species involved is the basis for further considerations of the mass transfer taking place in the system. Mastery of the relations between the fundamental units and measures involved, such as mole fractions, ratios, and mass fractions, is an important first step in developing this capability.

This chapter is devoted not only to the development of the basic principles of carrying out material balances but also to the summarization of the different units and measures used in biomedical engineering problems. In addition, several examples of simple material balance problems found in physiology are offered.

An initial summary of the units and measures used in material balance computations will be found in Table 2.1.

GENERAL MATERIAL BALANCE EQUATIONS

A general statement for the conservation of any species i in any given system may be written as follows:

$$\begin{matrix}\text{Accumulation of } i \\ \text{in the system}\end{matrix} = \begin{matrix}\text{the sum of} \\ \text{all inputs}\end{matrix} - \begin{matrix}\text{the sum of} \\ \text{all outputs}\end{matrix} + \begin{matrix}\text{net rate of} \\ \text{generation} \\ \text{of } i \text{ within} \\ \text{the system}\end{matrix}$$

Table 2.1. Measurement Terms Used in Material Balances

Quantity	*Symbol*	*Relation*	*Sum**
Mass of species *i*	m_i		m_{tot}
Moles of species *i*	n_i		n_{tot}
Mass fraction of *i*	ω_i	m_i/m_{tot}	1
Mole fraction of *i*	x_i, y_i†	n_i/n_{tot}	1
Partial volume of *i*	V_i		V
Mass density of *i*	ρ_i	m_i/V	ρ
Molar density of *i*	C_i	n_i/V	C
Overall mass density	ρ	m_{tot}/V	
Overall molar density	C	n_{tot}/V	
Specific volume of *i*	$\hat{V}_i$	V_i/m_i	
Specific molar volume	$\tilde{V}_i$	V_i/n_i	
Mass ratio of *i*	Ω_i	$m_i/m_{tot} - m_i$	
Mole ratio of *i*	X_i, Y_i	$n_i/n_{tot} - n_i$	

*Summed over all *i*, that is, totaling all the components.
†*x*'s are used for liquid phase, *y*'s for gas phase.

This statement, which might be called the principle of conservation, may be written in algebraic form as

$$dm_i/dt = \sum_{\text{in}} w_i - \sum_{\text{out}} w_i + \dot{r}_i \tag{2.1}$$

where

m_i = the total amount of *i* in the system, *M*

t = time (t)

$\sum_{\text{in}} w_i$ = the sum of all mass flows of *i* into the system, M/t

$\sum_{\text{out}} w_i$ = the sum of all mass flows of *i* out of the system, M/t

$\dot{r}_i$ = the net rate at which *i* is produced within the system, M/t (negative if *i* is being consumed)

The w_i terms in the above equation can be written as the product of the overall mass flow rate w and the mass fraction ω_i as

$$w_i = w\,\omega_i \tag{2.2}$$

When (2.1) is summed over all the species present, an overall mass balance results:

$$dm_{tot}/dt = \sum_{\text{in}} w - \sum_{\text{out}} w \tag{2.3}$$

with the symbols defined as in Table 2.1. Note that the sum of the r_i terms must equal zero, since in the absence of relativistic effects no mass can be created or destroyed.

The molal equivalents of these balances may be written as follows:

$$dn_i/dt = \sum_{\text{in}} \dot{N}_i - \sum_{\text{out}} \dot{N}_i + \dot{R}_i \tag{2.4}$$

where

n_i = the total number of moles if i in the system

$\dot{N}_i$ = the molal flow rates into and out of the system of i

$\dot{R}_i$ = the net rate at which species i is produced or consumed within the system, moles/time

As in the mass case each $\dot{N}_i$ can be written as the product of the overall molar flow rate and the mole fraction x_i as

$$\dot{N}_i = \dot{N} x_i \tag{2.5}$$

Unlike the mass equivalent, when the individual mole balances for each species are summed over all species, the generation term does not necessarily disappear, since moles are not necessarily conserved during chemical reactions. The following results:

$$dn_{\text{tot}}/dt = \sum_{\text{in}} \dot{N} - \sum_{\text{out}} \dot{N} + \sum_{i} \dot{R}_i \tag{2.6}$$

For that reason (2.6) is not of as much general utility as (2.3).

For *isochoric systems* (constant volume) these material balance relations can be written in more usable form, using volumetric flow rates rather than mass or molar rates. The following result:

Mass:

$$V\,(d\rho_i/dt) = \sum_{\text{in}} Q\,\rho_i - \sum_{\text{out}} Q\,\rho_i + \dot{r}_i \tag{2.7}$$

Moles:

$$V\,(dC_i/dt) = \sum_{\text{in}} Q\,C_i - \sum_{\text{out}} Q\,C_i + \dot{R}_i \tag{2.8}$$

where Q = the volumetric flow note, L^3/t.

If the processes taking place in the system are at *steady state*—that is, there is no change with time within the system—steady-state forms of the above two balances may be written as follows:

Mass:

$$\sum_{\text{in}} w\,\omega_i = \sum_{\text{out}} w\,\omega_i - \dot{r}_i \tag{2.9}$$

$$\sum_{\text{in}} Q\,\rho_i = \sum_{\text{out}} Q\,\rho_i - \dot{r}_i \tag{2.10}$$

Moles:

$$\sum_{\text{in}} \dot{N}\,x_i = \sum_{\text{out}} \dot{N}\,x_i - \dot{R}_i \tag{2.11}$$

$$\sum_{\text{in}} Q\,C_i = \sum_{\text{out}} Q\,C_i - \dot{R}_i \tag{2.12}$$

It is useful at this point to recall some relationships for perfect gases, the main basis being the ideal gas law

$$P\,V = n\,R\,T \tag{2.13}$$

which may be written alternatively as

$$P\,\widetilde{V} = R\,T \tag{2.14}$$

or

$$p_i\,V = n_i\,R\,T \tag{2.15}$$

or

$$P\,V_i = n_i\,R\,T \tag{2.16}$$

since

$$p_i/P = V_i/V = n_i/n \tag{2.17}$$

and

$$\sum_i p_i = P$$

$$\sum_i V_i = V$$

$$\sum_i n_i = n \tag{2.18}$$

Equations (2.17) and (2.18) are merely statements of Dalton's and Amagat's laws, while (2.15) and (2.16) define the partial pressure and partial volume of a species i in a mixture of gases.

It is also useful here to introduce the concept of *mole ratios*, which differ distinctly from mole fractions and are of great use in simplifying material balance calculations. The mole ratio of a species i, Y_i in a mixture is defined as

For gases:

$$Y_i = n_i/(n - n_i) = y_i/(1 - y_i)$$

$$= \text{moles of } i/\text{moles of mixture free of } i \qquad (2.19)$$

For liquids:

$$X_i = x_i/(1 - x_i) \qquad (2.20)$$

Example 2.1. Material Balance around a Normal Lung

In this example some of the relations given above will be used to analyze what happens when a 500 ml breath of fresh air is inspired into a normal healthy adult lung. All values of the quantity of each species present will be converted to their volumetric equivalents at a normal body temperature of 37° C and a pressure of 1 atm (BTP). The system considered will be the gas spaces in the lungs and its airways. Figure 2.1 schematically illustrates the system, with all the

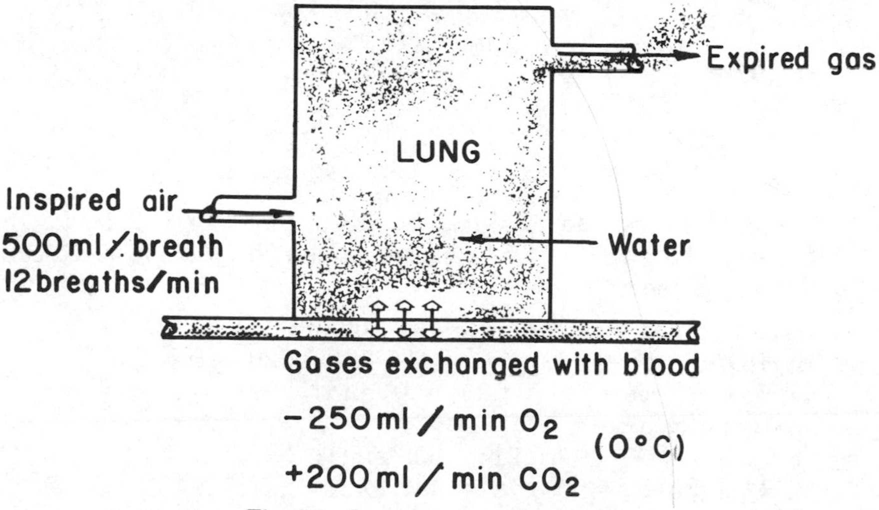

Fig. 2.1. Lung inputs and outputs.

input and output streams indicated. The problem is somewhat complicated by the fact that the inspired gas will pick up water and will also exchange oxygen for carbon dioxide as it passes through the system. We will assume a standard resting metabolic load of 250 ml/min of oxygen consumed and 200 ml/min of CO_2 produced. Since these quantities are usually measured and reported at standard temperature and pressure, (0° C instead of 37° C) (STP) they must be converted to BTP. They then become 284 and 226 ml/min respectively.

We will follow four significant molecular species in the gas phase and assume that any other species will be present only in negligible quantity. The four species to be followed are nitrogen, oxygen, carbon dioxide, and water.

Since no chemical reactions take place in the gas phase, the steady-state molar balance without the source or reaction term will be used. Equation (2.11) may then be written:

$$\sum_{\text{in}} \dot{N}_i = \sum_{\text{out}} \dot{N}_i \tag{2.21}$$

For one breath, converting the molar flow rates to the volumetric equivalent contained in one breath,

$$\sum_{\text{in}} V_i = \sum_{\text{out}} V_i \tag{2.22}$$

By consulting Figure 2.1, (2.22) may be written for each component as

$$V_{O_2 \text{ inspired}} = V_{O_2 \text{ expired}} + V_{O_2 \text{ lost to blood}} \tag{2.23a}$$

$$V_{CO_2 \text{ inspired}} = V_{CO_2 \text{ expired}} - V_{CO_2 \text{ gained from blood}} \tag{2.23b}$$

$$V_{N_2 \text{ inspired}} = V_{N_2 \text{ expired}} \tag{2.23c}$$

$$V_{H_2O \text{ inspired}} = V_{H_2O \text{ expired}} - V_{H_2O \text{ absorbed from lungs}} \tag{2.23d}$$

The partial pressure of water in the inspired air will be taken arbitrarily as 15 mm Hg. The relative amounts that are inspired can then be computed as follows:

Total volume of wet air inspired = 500 ml
Ml H_2O/ml wet air = p_{H_2O}/P = 15/760 = 0.020
Ml H_2O inspired = 500(0.020) = 10 ml H_2O
Ml dry air inspired = 500 - 10 = 490
Ml O_2 inspired = 490(0.21) = 103 ml O_2
Ml N_2 inspired = 490(0.79) = 387 ml N_2
Ml CO_2 inspired = 0 ml CO_2 (neglecting CO_2 in air)

The composition of the expired breath, assuming that it is completely saturated with water, can then be computed using (2.23) and Figure 2.1.

If the wet air leaving the lungs is completely saturated with water, its partial pressure will be equal to the vapor pressure of water at 37° C, which is 47 mm Hg. Then, the following computations result:

Ml H_2O/ml dry gas leaving = 47/(760 - 47) = 0.066
Ml O_2 expired = 103 - (284 ml/min)/(12 breaths/min) =
78 mls O_2

Ml N_2 expired = 387 ml N_2
Ml CO_2 expired = 0 + (226 ml/min)/(12 breaths/min) =
19 mls CO_2
Ml dry gas expired = 78 + 387 + 19 = 484 ml
Ml H_2O expired = 484(0.066) = 32 ml H_2O
Total ml expired = 516 ml

The following points should be noted about the computations:

1. The amount of dry gas leaving had to be computed before the amount of water leaving could be computed, since the total amount of wet gas was unknown. To complete the computation we made use of the *mole ratio* concept, rather than using mole fraction for the water calculation. The particular mole ratio employed, ml H_2O/ml dry air—or alternately the moles of H_2O/mole of dry air—is called the *humidity* and is extremely useful in air-water calculations.
2. The equivalent volume of gas leaving is greater than the amount entering, due to the increase in water content and due also to the fact that more CO_2 is gained than entered. This difference implies that the expired gas must be subjected to a higher pressure than the inspired gas, which is indeed the case. The ratio of volumes, 516/500 or 1.03, agrees well with observed values for expiratory versus inspiratory pressures, a typical example being, in mm of Hg, (760 + 20)/(760 − 10) = 780/750 = 1.04.

A summary of the results obtained above is shown in Table 2.2.

Table 2.2. Results of Overall Lung Material Balance

Species	*Inspired Gas*		*Expired Gas*		*Net Change*	
	ml	*%*	*ml*	*%*	*ml*	*%*
O_2	103	20.6	78	15.1	−25	−5.5
CO_2	0	0	19	3.7	+19	+3.7
N_2	387	77.4	387	75.0	0	−2.4
H_2O	10	2.0	32	6.2	+22	+4.2
Totals	500	100.0	516	100.0	+16	

Next, it is instructive for this particular problem, both on physiological and material balance grounds, to divide the system into two subsystems and to repeat the computation for each. We shall consider the lung in two parts, a dead space where no O_2 or CO_2 transfer takes place and the rest of the lung, which we will designate as the alveolar space. The dead space will be assigned a volume of 150 ml. Figure 2.2 depicts the new system.

Equations (2.23a–d) may be rewritten for each subsystem. The only significant change will be that the dead space equations will

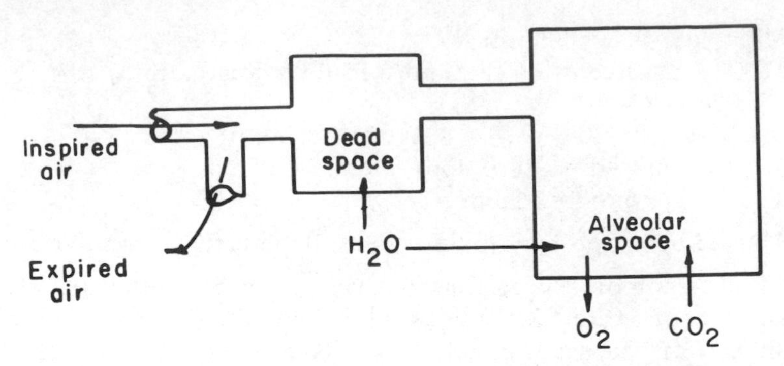

Fig. 2.2. Lung inputs and outputs with dead space.

allow for water transfer but for no O_2 or CO_2 transfer. Also, we note that as shown in Figure 2.2 the expired air leaves through the same dead space as the inspired air entered—a much more reasonable scheme. The computation now proceeds as follows.

Dead space—in:
Ml into dead space = 150 ml
Ml H_2O inspired = 150(0.02) = 3 ml H_2O
Ml N_2 inspired = 147(0.79) = 116 ml N_2
Ml CO_2 inspired = 0 ml CO_2
Ml O_2 inspired = 147(0.21) = 31 ml O_2

This 150 ml will push 150 ml of alveolar air that was left in the dead space from a previous expiration back into the lungs. Subsequently, the fresh air is pushed back out, having picked up water. The rest of the species will stay the same.

We may then compute the composition of the expired gases leaving the dead space as follows:

Dead space—out:
Ml H_2O out of dead space = 147(0.066) = 9 ml H_2O
Ml O_2 out = 31 ml O_2
Ml CO_2 out = 0 ml CO_2
Ml N_2 out = 116 ml N_2
Total ml out = 9 + 31 + 116 + 0 = 156 ml

Since we have previously determined the total amount of each species expired and from the previous calculation we have determined what came from the dead space, we can now compute what comes from the alveolar space; and further, if we stipulate that what comes from the alveolar space is representative of the contents of that space, we can compute the alveolar composition.

Alveolar space:

Ml H_2O from alveoli = 32 - 9 = 23 ml H_2O
Ml O_2 from alveoli = 78 - 31 = 47 ml O_2
Ml CO_2 from alveoli = 19 - 0 = 19 ml CO_2
Ml N_2 from alveoli = 387 - 116 = 271 ml N_2
Total ml from alveoli = 23 + 47 + 19 + 271 = 516 - 156 = 360 ml

A summary of these results is presented in Table 2.3. Percentage compositions are given both on a wet and a dry basis.

Table 2.3. Summary of Lung Concentrations

Basis: 500-ml breaths, 150 ml dead space, $p_{H_2O\ in}$ = 15 mm
$\dot{V}_{O_2}$ = 250 ml/min, R = 0.8

Species	*Inspired Gas*		*Alveolar Gas*		*Expired Breath*	
	Wet	*Dry*	*Wet*	*Dry*	*Wet*	*Dry*
H_2O	2.0%	. . .	6.2%	. . .	6.2%	. . .
O_2	20.6	21.0	13.1	14.0	15.1	16.0
CO_2	0.0	0.0	5.3	5.7	3.7	4.0
N_2	77.4	79.0	75.4	80.3	75.0	79.6

The partial pressures of CO_2 and O_2 in the alveolar space can be computed from Table 2.3, on either a wet basis or dry basis as follows:

Dry basis:

$$p_{O_2} = 0.140(760 - 47) = 100 \text{ mm Hg}$$

$$p_{CO_2} = 0.057(760 - 47) = 40 \text{ mm Hg}$$

Wet basis:

$$p_{O_2} = 0.131(760) = 100 \text{ mm Hg}$$

$$p_{CO_2} = 0.053(760) = 40 \text{ mm Hg}$$

As they must be, these values are independent of the basis on which they were calculated. They are also representative of the values usually given as typical for this situation. It should also be pointed out here that many authors of physiology texts use sets of alveolar ventilation rates, metabolic rates, and alveolar partial pressures which are not consistent with material balance considerations, as can be shown by the type of material balance computations illustrated here.

The manipulations leading to the results of Table 2.3 could be rearranged in part to give some general relations relating dead space volume V_D, tidal volume V_T, and the relative concentrations F_i for all the species except water on a dry basis as follows:

$$V_D/V_T = (F_E - F_A)/(F_I - F_A) \tag{2.24}$$

or

$$F_A = (V_T F_E - V_D F_I)/V_T - V_D) \tag{2.25}$$

where the subscripts T, D, E, I, and A denote tidal, dead space, expired, inspired, and alveolar respectively.

An interesting by-product of this example is the computation of the net rate at which mass is lost through the respiratory process by the body during an average day. The difference between the CO_2 lost and the O_2 gained, and the net rate at which water is lost can be computed directly from the previous results using the ideal gas law.

$$O_2 \text{ gained/min} = \frac{250 \text{ ml}}{\text{min}} \left| \frac{\text{mole}}{22{,}400 \text{ ml}} \right| \frac{32 \text{ gm}}{\text{mole}} = 0.315 \text{ gm/min}$$

$$CO_2 \text{ lost/min} = \frac{200 \text{ ml}}{\text{min}} \left| \frac{\text{mole}}{22{,}400 \text{ ml}} \right| \frac{44 \text{ gm}}{\text{mole}} = 0.345 \text{ gm/min}$$

$$H_2O \text{ lost/min} = \frac{264 \text{ ml}}{\text{min}} \left| \frac{\text{mole}}{22{,}400 \text{ ml}} \right| \frac{18 \text{ gm}}{\text{mole}} \left| \frac{273^\circ \text{K}}{310^\circ \text{K}} \right.$$

$$= 0.187 \text{ gm/min}$$

$$\text{Net loss} = -0.217 \text{ gm/min}$$

$$\text{Net rate of mass lost per day} = 0.217 \text{ gm/min} \times 1440 \text{ min/day} = 313 \text{ gm/day}$$

In other words, a weight loss of nearly three-fourths of a pound per day can be experienced just by normal breathing. The energy losses associated with this process were previously estimated in Example 1.7 of Chapter 1.

This example instructively can be extended further to include the blood entering and leaving the lungs. Figure 2.3 depicts the revised system that will be considered. This time we will change the basis from one breath to one minute and make use of the results already computed. Equation 2.12 is the appropriate form to use here.

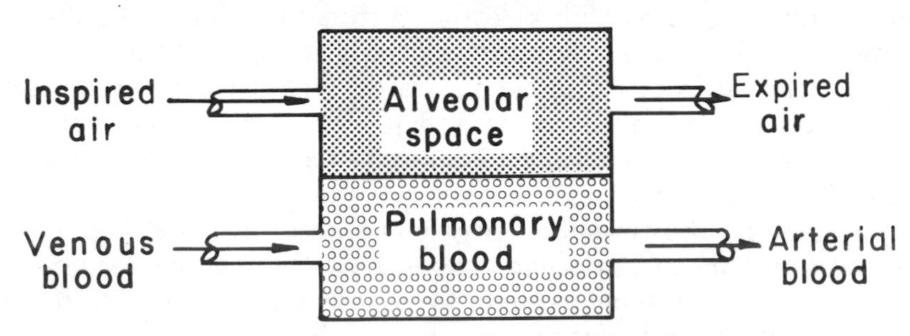

Fig. 2.3. Alveolar-blood gas exchange.

Although the O_2 and CO_2 unquestionably participate in chemical reactions within the system, specifically in the red cells, we measure their equivalent concentrations in blood as though they were merely dissolved rather than bound species. We speak of a typical arterial oxygen content of blood, for example, as C_{O_2} = 20 ml O_2/100 ml blood and understand this implies that if all the oxygen present in all forms (dissolved in solution as well as bound chemically) in 100 ml of blood were completely released, it would occupy a volume of 20 ml when measured at a pressure of 1 atm and a temperature of 37° C.

Looking at the system from this standpoint and also considering that no new O_2 or CO_2 is created or destroyed within the system, the reaction term R_i in (2.12) may be dropped. Next, we convert the results of the previous calculation into flow rates and concentrations as follows:

$$
\begin{aligned}
Q_{\text{gas out}} &= 516 \text{ ml/breath} \times 12 \text{ breaths/min} = 6{,}192 \text{ ml/min} \\
Q_{\text{gas in}} &= 500 \text{ ml/breath} \times 12 \text{ breaths/min} = 6{,}000 \text{ ml/min} \\
C_{O_2\,\text{in}} &= 0.206 \text{ ml } O_2\text{/ml gas} \\
C_{CO_2\,\text{in}} &= 0.00 \text{ ml/ml gas ml } CO_2\text{/ml gas} \\
C_{O_2\,\text{out}} &= 0.151 \text{ ml } O_2\text{/ml gas} \\
C_{CO_2\,\text{out}} &= 0.037 \text{ ml/ml gas ml } CO_2\text{/ml gas}
\end{aligned}
$$

Choosing a representative value of the blood flow rate of

$$Q_B = 5{,}000 \text{ ml/min}$$

and specifying typical values of the venous blood concentrations as $C_{B\,O_2\,\text{in}} = 0.145$ and $C_{B\,CO_2\,\text{in}} = 0.520$, we can demonstrate the use of (2.12) and compute the resulting arterial blood concentrations (subscript B indicating blood).

For O_2:

$$Q_{\text{gas in}}\, C_{O_2\,\text{in}} + Q_B\, C_{B\,O_2\,\text{in}} = Q_{\text{gas out}}\, C_{O_2\,\text{out}} + Q_B\, C_{B\,O_2\,\text{out}}$$

$$C_{B\,O_2\,\text{out}} = \frac{6{,}000(0.206) - 6{,}192(0.151)}{5{,}000} + 0.145 = 0.195$$

For CO_2:

$$C_{B\,CO_2} = 0.480 \text{ (details left for the reader)}$$

These concentration values correspond to the partial pressure values computed earlier. That is, blood with the oxygen and carbon dioxide contents as computed above is in chemical equilibrium with a gas having partial pressures of oxygen and carbon dioxide as computed earlier, of 100 and 40 mm Hg respectively. This area of gas-liquid equilibrium will be treated in depth in succeeding chapters.

This concludes our treatment for the time being of this example. It should be noted that a great deal of information can be obtained by the simple and consistent application of basic material balance principles.

Example 2.2. Material Balance around an Artificial Kidney

The facility of working with different sets of units is an important aspect of material balance computations. In this example additional common concentration units used in physiology are introduced, and their use is demonstrated in a problem involving the hemodialysis procedure used to supplement or replace human kidney function.

The concentration units commonly used to denote plasma and urine concentrations are defined as follows:

Mg % = mg of the species/100 ml of solution

$$\text{(Meq)/liter} = \frac{\text{mass of species in mg} \times \text{valence of species}}{\text{volume of solution in liters} \times \text{molecular weight of the compound}}$$

Example:

0.25 gm of H_3PO_4 in 2 liters of solution

Meq/liter of H^+ = 250(3)/98(2) = 3.8 meq/liter

Mg % of H_3PO_4 = 250/20 = 12.5 mg %

The artificial kidney system to be considered in this example is illustrated in Figure 2.4.

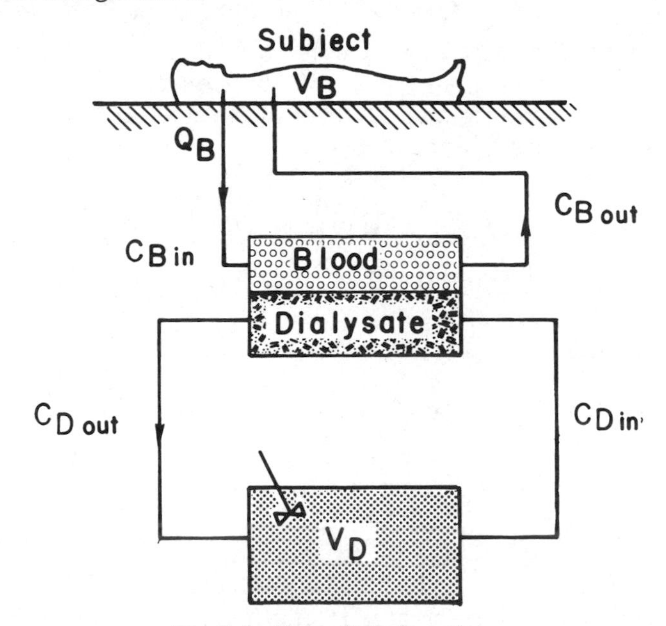

Fig. 2.4. Hemodialysis system.

The following data are available:

Initial PO_4 blood level	= 110 meq/liter	Q_B	= 150 ml/min
Desired final PO_4 blood level	= 20 meq/liter	Q_D	= 1,500 ml/min
Initial urea level	= 3,900 mg %	V_B	= 5,000 ml
Desired final urea level	= 1,200 mg %	V_D	= 50 liters

Initially the dialysate fluid will be free of both urea and phosphate. The following items are required:

1. The total amount of PO_4 and urea to be removed:

   Meq of PO_4 removed = 5 liters (110 meq/liter - 20 meq/liter)
   = 450 meq

   Mg of urea removed = 5 liters (3,900 - 1200)(mg %) (10)
   × (10^{-3} gm/mg)
   = 135 gm

2. The final dialysate concentration. Since all the substances removed end up in the dialysate phase:

   Final PO_4 concentration = 450 meq/50 liters = 9 meq/liter
   Final urea concentration = 135 gm/50 liters = 2.7 gm/liter
   = 270 mg %

3. At any moment during the procedure (2.12) may be applied to relate the compositions of the four streams. For example, at the beginning of the procedure when the initial blood concentrations apply:

$$Q_B\ (110 - C_{B\,\text{out}}) = Q_D\ (C_{D\,\text{out}} - 0)$$

or

$$C_{B\,\text{out}} = 110 - 10\ C_{D\,\text{out}} \tag{2.26}$$

It is obvious that another relationship between C_B and C_D will be required in order to solve for specific values of these quantities at any given time. This extra relation cannot be obtained from material balance considerations. It must come from a consideration of the rate at which the species in question can be transferred in the dialyzing device. The quantification of such rates is the major subject of Chapters 7 and 8. Equation 2.26 does, however, constrain the relative values of C_B and C_D in an exact fashion.

It is important to point out here that the use of an overall blood concentration such as C_B to characterize the content of such ions as PO_4 is an oversimplification which will lead to significant error in an analysis of the rates of material transfer in systems such as these. A more correct formulation would involve separately accounting for the species contained in the plasma phase and in the red cell phase as follows:

$$C_B = (1 - H)\, C_P + H\, C_R \qquad (2.27)$$

where

H = volume fraction red cells present
C_P = concentration in the plasma phase
C_R = concentration in the red cell phase

For the purpose of material balance computations, the use of overall concentrations is acceptable if the proper care is taken to account for all streams present.

Example 2.3. Overall Daily Water Balance

The daily water balance for a normal individual is an important example of an unsteady state or transient material balance problem. We may begin the formulation of this problem by considering the separate terms in the general overall statement of the principle of conservation.

Accumulation = sum of inputs - sum of outputs
+ net amount produced

See (2.1).

For body water, every term in this equation is significant. We shall drop the subscript i, understanding that in the balance of this example we are referring to water, and discuss each term in the equation separately.

1. The term dm_i/dt represents the net rate at which the body accumulates or loses water, measured at any given time. It is a differential quantity and has the units of mass of water per unit time.
2. The term $\Sigma_{in}\, w$ represents the sum of all the inputs of water to the body. Normally there are three:
   a. Water physically bound to ingested food = w_f.
   b. Water ingested as liquid = w_d.
   c. Water inspired as vapor in breathing = w_I.
3. The term $\Sigma_{out}\, w$ represents the sum of all the outputs of water from the body. Normally there are five:
   a. Water lost as vapor in expired breath = w_e.
   b. Water lost as vapor from exposed skin = w_v.
   c. Water lost as liquid from sweat runoff = w_s.
   d. Water physically bound to feces = w_w.
   e. Water lost in the urine = w_u.

In some abnormal situations significant water is lost in regurgitation or may be gained by transfusion.

4. The last term $\dot{r}_i$ represents a significant gain of water for the body. It represents the water that is produced by chemical reaction resulting mostly from the oxidation of foodstuffs. For example, in the overall conversion of carbohydrates (such as glucose)

$$C_6H_{12}O_6 + 6\ O_2 \longrightarrow 6\ CO_2 + 6\ H_2O \qquad (2.28)$$

six moles of water are formed for each mole of glucose burned.

Equation (2.1) may then be written for water as

$$dm/dt = w_f + w_d + w_I - w_e - w_v - w_s - w_w - w_u + \dot{r} \qquad (2.29)$$

This gives the relationship between the instantaneous rates of each of these inputs and outputs. The relationship between the total amounts of water over some period of time is usually of more interest. This must be obtained from the integrated form of (2.29), which follows the general principle that

$$\Delta m = \int_t^{t+\Delta t} (dm/dt)\, dt \qquad (2.30)$$

Since many of the inputs and outputs are not smooth or continuous functions of time, the integration is usually not done analytically, but by summation of separate terms. In either case, the integrated form may be written as

$$\Delta m = m_f + m_d + m_I - m_e - m_v - m_s - m_w - m_u + \bar{r} \qquad (2.31)$$

where

$$m_f = \int_t^{t+\Delta t} w_f\, dt = \text{total amount of water in ingested food, etc.}$$

$$\bar{r} = \int_t^{t+\Delta t} \dot{r}\, dt \qquad (2.32)$$

Typical values for these terms over a time period of 24 hours might be:

In		*Out*	
m_f	= 1,000 gm	m_e	= 440 gm
m_d	= 2,000 gm	m_v	= 700 gm
m_I	= 40 gm	m_w	= 150 gm
$\bar{r}$	= 400 gm	m_u	= 1,500 gm
	3,440 gm	m_s	= 650 gm
			3,440 gm

$\Delta m = 0$

Equation (2.31) can be used to estimate unbalance over any time period for which (2.30) can be directly integrated. Equation (2.29) in turn can be used when information about the rates of each stream are desired.

Example 2.4. Measurement of Blood Plasma Volume by Dye Dilution

One standard and historic method of measuring the amount of certain body fluids present in the system is the use of a measured amount of dye that is preferentially soluble in the fluid to be measured. A measured amount of dye of known concentration is injected into the circulatory system for example, and after a certain length of time the concentration in the plasma is measured and the systemic plasma volume is computed. The so-called "Fick principle" is applied to compute the result. In this example we will show that the Fick principle is simply a special form of the more general expression (2.8), the general material balance for constant volume systems.

For one input stream, with no outputs and no chemical reaction, (2.8) may be written as

$$V\,dC/dt = Q_{in}\,C_{in} \tag{2.33}$$

Multiplying through by dt

$$V\,dC = Q_{in}\,C_{in}\,dt \tag{2.34}$$

If the volume change due to dye addition is negligible compared to the systemic volume, this relation may be integrated over the time of injection,

$$\int_{C_o}^{C_f} V\,dC = \int_0^{t_o} Q_{in}\,C_{in}\,dt \tag{2.35}$$

to give

$$V\,(C_f - C_o) = V_{in}\,C_{in} = \text{mass of dye injected} \tag{2.36}$$

which may be rearranged to give

$$V = V_{in}\,C_{in}/(C_f - C_o) \tag{2.37}$$

where

V = system volume (usually to be determined)
C_f = final concentration of solute dye
C_o = initial concentration of solute dye (usually zero)
V_{in} = volume of injected solution
C_{in} = concentration of solute dye in injected solution

If the volume of added solution is significant, then the reader may verify that the correct solution is

$$V = V_{in}\,(C_{in} - C_f)/(C_f - C_o) \tag{2.38}$$

This result may be obtained from integration of (2.1) or more simply from a direct overall material balance.

An effect which must be taken into account here is the relatively slow absorption by the rest of the system of the dye dissolved in that part of the system to be measured. Although such dyes are usually relatively "nondiffusible," this effect must be watched for.

Dye dilution measurements in the circulatory system are further complicated by the difficulty of getting a representative sample to measure C_f, and also by the effects of repeated recirculation. Fortunately, the circulation phenomena allows another important piece of information to be obtained from this type of measurements—the mean flow rate in the system and the cardiac output in the case of the circulatory system.

Figure 2.5 shows a typical concentration versus time record ob-

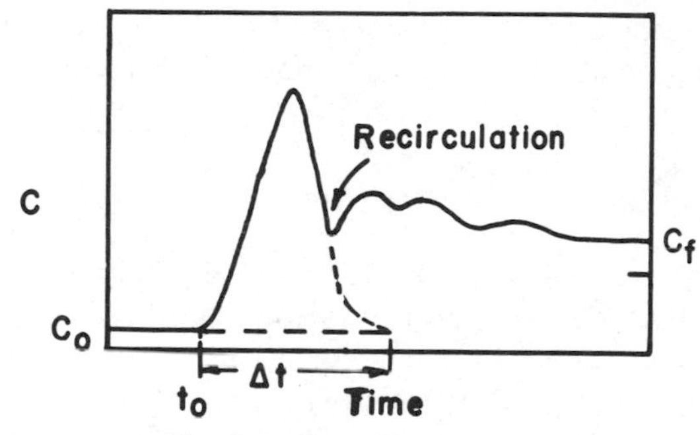

Fig. 2.5. Dye dilution data.

tained in the right ventricle. The effects of recirculation can be plainly seen as the blood coming through for the second time adds more dye per unit volume to the sampling region. After several recirculations, the concentration levels off to the final value predicted by (2.37).

The information needed to compute the cardiac output is all contained in the first peak of Figure 2.5. It is necessary to extrapolate the response to cancel out the effects of recirculation. One way is to plot the data of Figure 2.5 on semilogarithmic paper, extrapolate the resulting straight line, and then replot the data on arithmetic paper to produce the dashed line shown in Figure 2.5.

Assuming that all the injected dye passes the sampling point (a good assumption in the case of the right ventricle) we can reintegrate the material balance equation as

$$\text{Mass of dye injected} = \int_0^{t_s} Q_{\text{in}} C_{\text{in}} \, dt = \int_{t_o}^{t_o+\Delta t} Q\, C\, dt \qquad (2.39)$$

where

t_s = time to inject the sample
t_o = the time to make one pass, as indicated in Figure 2.5
Δt = width of the extrapolated peak
Q = systemic flow rate, L^3/t
C = dye concentration at the sampling point

The second integral in (2.39) can be rewritten as $Q \int_{t_o}^{t_o+\Delta t} C\,dt$ if the flow rate is independent of time. The concentration integral can now be obtained by graphical integration (planimetry) of the extrapolated peak, and the flow rate, or cardiac output, can be obtained by

$$Q = \text{mass of dye injected/integral} \qquad (2.40)$$

In this example we have seen how two pieces of information, the total system volume and the volumetric flow rate, can be obtained by simply applying material balance concepts to the results of the experiment.

SUMMARY

This chapter has dealt with the basic equations and techniques used for material balance computations and has presented some examples demonstrating their use for a variety of situations. While simple in concept and application, these techniques are extremely important in paving the way for the consideration of energy and mass transfer to be discussed in the following chapters.

Bibliography

Many good textbooks on material balance techniques are available. Among these are:

Himmelblau, D. L. *Basic Principles and Calculations in Chemical Engineering*. Prentice-Hall, 1961.

Hougen, O. A., Watson, K. M., and Ragatz, R. A. *Chemical Process Principles*. Wiley, 1943.

Background material on the water balance and kidney calculations is available in:

Pitts, R. F. *Physiology of the Kidney and Body Fluids*. Year Book Medical Publishers, 1963.

Background material for the material balances on the respiratory system can be found in:

Comroe, J. H., Jr. *Physiology of Respiration*. Year Book Medical Publishers, 1965.

Additional information on dye-dilution methods for flow and volume measurements is obtainable from:

Wood, E. H., ed. *Symposium on the Use of Indicator Dilution Techniques in the Study of Circulation.* American Heart Association, 1962.

Two other interesting applications of the general techniques discussed in this chapter are found in:

Bischoff, K. B., and Brown, R. G. Drug distribution in mammals, pp. 32–44. *Chemical Engineering Progress Symposium Series,* vol. 62, no. 66, American Institute of Chemical Engineers, 1966.

Spencer, J. L., Kinney, J. M., and Long, C. L. Material and energy balances on post-operative patients, pp. 123–30. *Chemical Engineering Progress Symposium Series,* vol. 62, no. 66, American Institute of Chemical Engineers, 1966.

chapter 3

Energy Balances in Closed Systems

INTRODUCTION

Equipped with a basic understanding of the modes of interconversion of energy from Chapter 1 and the basic material balance formulations of Chapter 2, we can next proceed to the consideration of energy balances in living systems. In this chapter, we shall consider energy relationships in systems we can regard for practical purposes as closed; that is, the effects of the material that crosses the system boundaries can either be neglected for the purpose of energy considerations, or they can be accounted for in another way while still employing closed-system thermodynamics. For example, the energy losses resulting from the evaporation of water through the body surfaces—clearly an example of material crossing the system boundaries—can be accounted for in a heat transfer term. In Chapter 4 we shall reexamine some of these terms in a more rigorous fashion.

METABOLIC ENERGY TRANSFER

We shall begin by investigating energy considerations for humans over relatively short intervals in between periods of food intake. The purpose of this example is not only to develop some useful relations but also to increase our understanding of the fundamental aspects of the energy transformations involved.

By temporarily considering the human body as a closed system, we are not only assuming that the mass is constant but we are also ignoring the oxygen and carbon dioxide exchanged with the atmosphere through the respiratory process. It is important to note that for energy consideration these streams can be neglected with very little error. Perhaps the simplest way to illustrate this would be to compute the enthalpy of these streams and to choose the reference temperature as the body temperature. In Chapter 4 this will be discussed in more detail. It will suffice at this point to state that any energy required to heat or cool the respiratory gases can be accounted for as a heat loss, as indicated above.

The closed-system energy balance equation (1.13) can be written for the human body as follows:

$$dU/dt + dK/dt + d\phi/dt = \dot{Q} - \dot{W} \tag{3.1}$$

where

U = total internal energy of the system
K = total kinetic energy of the system
ϕ = total potential energy of the system
$\dot{Q}$ = energy transferred to the system as heat
$\dot{W}$ = work done by the system

For almost every situation of interest the rate of change of potential and kinetic energy of a living system will be very much less than the other terms in (3.1). Also, most living systems will most of the time lose heat to the surroundings rather than gain it, since they operate at a temperature higher than that of the environment. Therefore, it is convenient to define

$$\dot{Q}' = -\dot{Q} \tag{3.2}$$

where $\dot{Q}'$ will represent energy lost to the surroundings as heat. We may then write

$$dU/dt = -\dot{Q}' - \dot{W} \tag{3.3}$$

This indicates that, at least between periods of food ingestion, living systems must be "running down," since the right side of this equation is almost always negative. This is, of course, exactly the case since it is only by the energy obtained from higher energy food intake that life is sustained.

We can arbitrarily separate the internal energy of the system into two parts—that associated with *temperature* and that associated with *composition*, that is, the chemical make-up. We have seen in Chapter 1 that certain compounds having higher enthalpies of formation have correspondingly higher relative internal energies at a given temperature. For example, a complicated molecule such as glucose has a much higher amount of internal energy stored in its bonds than does an equivalent amount of carbon dioxide.

Then we can write the internal energy as

$$U = U_T + U_C \tag{3.4}$$

and denoting

$$dU/dt = dU_T/dt + dU_C/dt = \dot{T} + \dot{M} \tag{3.5}$$

from Chapter 1 we remember that

$$\dot{T} = dU_T/dt = \frac{d}{dt}(m_{tot}\hat{C}_p T) = (m_{tot}\hat{C}_p)\, dT/dt \tag{3.6}$$

where T is the average temperature of the system, in this case the mean body temperature. The average heat capacity of the system $\hat{C}_p$ is in this case about 0.86 kcal/kg° C.

The term $\dot{M}$ represents the rate at which stored chemical energy is converted to other forms, eventually to be lost across the system boundaries as heat or work. This overall balance tells us nothing about how the energy that is converted is used within the system (for example, to pump, synthesize, or transport), but only describes its relation to the overall systemic exchanges. $\dot{M}$ is usually termed the metabolic energy conversion rate, or more simply the metabolism.

Equation 3.3 may now be written

$$\dot{M} + \dot{T} = -\dot{Q}' - \dot{W} \tag{3.7}$$

An important special case of this expression describes a resting ($\dot{W} = 0$) system at constant temperature. Under these conditions we denote the values of $\dot{M}$ and $\dot{Q}'$ with the subscript o to indicate the basal or normal condition:

$$\dot{M}_o = -\dot{Q}'_o \tag{3.8}$$

For normal adult males $\dot{M}_o$ is approximately -70 kcal/hr and is called the basal metabolism. It can be measured either directly by measuring $\dot{Q}'_o$ in a calorimeter or indirectly by measuring the oxygen consumption rate and using the results described in Example 1.6 of Chapter 1 to compute $\dot{M}_o$. The latter method requires less sophisticated equipment and hence is more generally used. An oxygen consumption rate of 250 ml/min as used in Chapter 2 corresponds to an $\dot{M}_o$ of -72.5 kcal/hr when the averaged figures of the calorific equivalent as presented in Example 1.6 are used.

One consequence of (3.8) is the realization that the body must exchange a considerable amount of energy—enough to heat 2 gal of water 10° C—every hour in order to operate isothermally.

For convenience, we divide $\dot{M}$ into two parts (both negative between meals) and write

$$\dot{M} = \dot{M}_o + \Delta\dot{M} \tag{3.9}$$

so that (3.7) becomes

$$\dot{M}_o + \Delta\dot{M} + \dot{T} = -\dot{Q}' - \dot{W} \tag{3.10}$$

This is the basic equation that will now be used to describe the relationship between body temperature, heat loss, work and exercise, and metabolic energy conversion. Some examples are worth considering.

Example 3.1. Shivering

If the body is exposed to low temperature, the following sequence often occurs:

1. $\dot{Q}'$ increases because of increased temperature difference.
2. $\dot{T}$ becomes negative, as a result of (3.10), and the body temperature drops accordingly.
3. In response to this, shivering begins, causing an increase in $\Delta\dot{M}$ corresponding to an extra amount of metabolic energy conversion in the muscle layers, until this negative term becomes large enough to change $\dot{T}$ back to a positive term and to make up for the environmental increase in $\dot{Q}'$.
4. In this process some external work may be done, requiring further increase in the negative term $\Delta\dot{M}$.

This example introduces the important relationship between external work and increased metabolic conversion ($\dot{W}$ and $\Delta\dot{M}$)—a most complicated subject from a physical point of view. The efficiency of the muscles in performing external work, in terms of the energy conversion required, has been estimated as about 20%, that is

$$\Delta\dot{M} \approx -5\,\dot{W} \tag{3.11}$$

or about 5 times more chemical energy is converted than is actually usefully employed in performing work—the difference either being dissipated as heat or, as in the above example, used to accomplish internal heating.

Since the efficiency of the muscle system will vary according to the type of external work being done, the particular muscles involved, and the condition of the individual involved, a more general representation of (3.11) is desirable:

$$\Delta\dot{M} = -n\,\dot{W} \tag{3.12a}$$

or

$$-\dot{W} = \Delta\dot{M}/n = b\,\Delta\dot{M} \tag{3.12b}$$

which allows (3.10) to be written as

$$\dot{M}_o + (1-b)\,\Delta\dot{M} + \dot{T} = -\dot{Q}' \tag{3.13a}$$

or

$$\dot{M}_o + \dot{T} = -\dot{Q}' + (n-1)\,\dot{W} \tag{3.13b}$$

Example 3.2. Body Temperature and Heat Transfer

In this example we will compute the rate at which the body temperature would initially rise if heat transfer to the environment were not permitted—the adiabatic temperature rise of the body. This could be accomplished by immersing a subject in a water bath at 37° C. For this case (3.10) would initially read

$$\dot{M}_o + \dot{T} = 0 \tag{3.14}$$

or

$$m_{\text{tot}}\, \hat{C}_p\, dT/dt = -\dot{M}_o \tag{3.15}$$

Using the values previously offered for a 70-kgm subject,

$$dT/dt = \frac{70 \text{ kcal/hr}}{(0.86 \text{ kcal/kgm °C})\ 70 \text{ kgm}} = 1.15\text{°C/hr} \tag{3.16}$$

Example 3.3. Energy Conversion in the Heart

It is useful here to discuss the relationship of the internal use of stored chemical (or internal) energy in a subsystem to the dissipation of energy across its surface. An understanding of this relationship is vital to further understanding of energy balances in living systems. Let us choose the action of the heart as an example and perform an

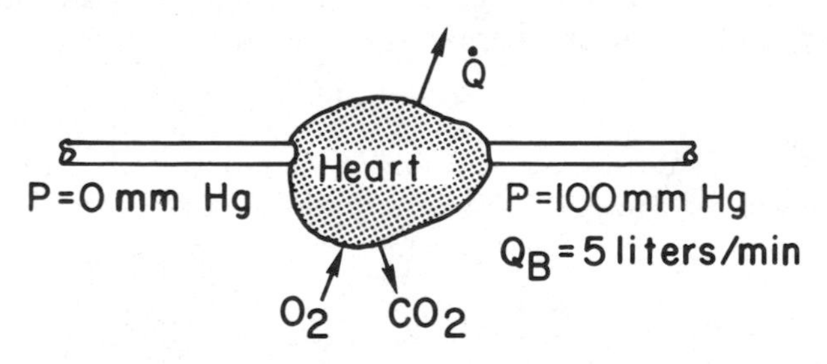

Fig. 3.1. Energy conversion in the heart.

energy balance around the subsystem of the heart, as indicated in Figure 3.1. An energy balance around this system will center on the evaluation of three terms:

1. The mechanical work done in pumping blood.
2. The conversion of stored chemical energy in the heart muscles to mechanical energy to accomplish 1.
3. The heat transferred away from the system as a result of the inefficiency in 2.

This may be summed up by the direct application of (3.7), assuming constant temperature.

$$\dot{M} = -\dot{Q}' - \dot{W} \tag{3.17}$$

In this case, the work performed, $\dot{W}$, can be computed directly since it is equal to the mechanical work performed in pumping blood

$$\dot{W} = Q_B \Delta P = \frac{5 \text{ liters/min } (100 \text{ mm Hg})}{51.7 \text{ mm Hg/psi}} \times \frac{0.154 \text{ milliHP}}{\text{psi liters/min}} \tag{3.18}$$

$$= 1.5 \text{ milliHP} = 23 \text{ kcal/day}$$

The $\dot{M}$ term can be computed by measuring the oxygen consumption of the cardiac muscles and then using the appropriate calorific oxygen equivalent, as developed in Example 1.6, of 4.825 kcal/liter of O_2 consumed.

$$\dot{M} \approx 5 \text{ ml } O_2/\text{min} \approx -35 \text{ kcal/day} \tag{3.19}$$

The energy lost as inefficiency and transferred away as heat directly from the subsystem is therefore about 35 - 23 = 12 kcal/day. These figures indicate, not surprisingly, that the heart is a very efficient converter of chemical energy, having an efficiency of about 65%.

All the fluid pumped by the heart remains within the body and eventually returns to the heart at the same flow rate and with its pressure or flow energy completely dissipated due to frictional losses in the circulatory system. Also, the body does not increase in temperature or internal energy at any point between meals. Then we see that all the 35 kcal/day converted in the heart muscles must eventually cross the system's boundaries primarily as heat. Another way to look at this is to redraw the system of Figure 3.2 and consider the entire cardiovascular system as the subsystem in question.

Obviously, this system does no mechanical work. It does transport oxygen and other nutrients, although on an overall basis the energy effects of these processes are extremely small. Everything that is picked up is dropped off somewhere else in the loop. Then (3.11) becomes

$$\dot{M} = -\dot{Q}' \tag{3.20}$$

as implied in the previous paragraph.

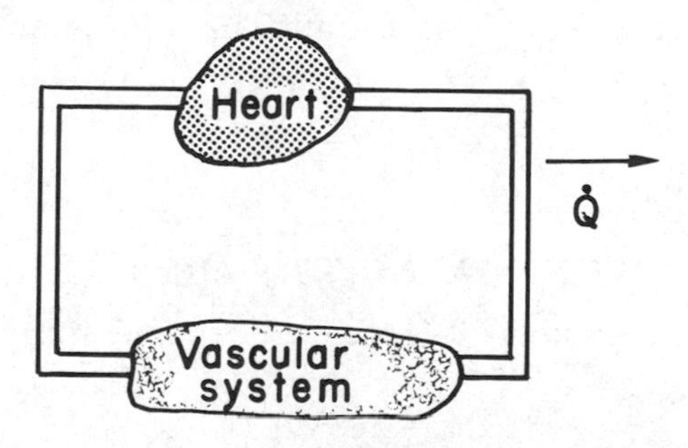

Fig. 3.2. Cardiovascular system.

An analysis of every subsystem of the body can be made in principle, with the resulting conclusion that each subsystem is a net user of stored chemical energy, either to do mechanical work as in the heart and muscles, to perform synthesis of high energy compounds as in the liver, to perform concentration processes as in the stomach acid glands (recalling Example 1.12), or to transport species against gradients as in the tubules of the nephron. The physical-chemical processes that accomplish these tasks vary in efficiency and in the modes by which they transform energy.

Example 3.4. ATP and Energy Storage

A commonly discussed example is the storage of energy in cells using the high-energy phosphate bond that differentiates adenosine diphosphate (ADP) from adenosine triphosphate (ATP). The overall chemical reaction that describes this may be written, using glucose as an example energy source, as

$$C_6H_{12}O_6 + 38\ PO_4 + 38\ ADP + 6\ O_2 \longrightarrow 6\ CO_2 + 38\ ATP + 44\ H_2O \tag{3.21}$$

The energy that can be stored in one phosphate bond is about 7 kcal/mole of ADP. The molecular weight of ADP is 473. The efficiency of this storage process can be estimated as

Efficiency of storage process

$$= \frac{\text{stored energy}}{\text{energy from glucose}}$$

$$= \frac{(38 \text{ moles ADP/mole glucose})\ (7 \text{ kcal/mole ADP})}{(673 \text{ kcal/mole glucose})}\ (100)$$

$$= \text{about } 40\% \tag{3.22}$$

This example indicates that this particular energy storage mechanism is relatively efficient.

At this juncture it is desirable to consider the relationship between stored chemical energy as internal energy and the concept of free energy. Recalling the basic definitions from Chapter 1,

$$G = H - T\,S \tag{3.23a}$$

$$H = U + P\,V \tag{3.23b}$$

then, on combination

$$dU = dG + d(TS) - d(PV) \tag{3.24}$$

For biological systems of constant temperature, pressure, and volume

$$dU = dG + TdS \tag{3.25}$$

or

$$dU/dt = dG/dt + T\, dS/dt \tag{3.26}$$

From the second law (1.36)

$$dS/dt = \dot{Q}/T + \dot{R}/T \tag{3.27}$$

Then

$$\dot{G} = dG/dt = dU/dt - \dot{Q} - \dot{R} \tag{3.28}$$

Since $Q = -Q'$ and using (3.5),

$$\dot{G} = \dot{M} + \dot{T} + \dot{Q}' - \dot{R} \tag{3.29}$$

so, from (3.7)

$$\dot{G} = -\dot{W} - \dot{R} \tag{3.30}$$

which states that the rate of decrease of free energy in the system is equal to the rate at which external work is done plus the rate at which internal energy is lost in irreversibilities such as friction. If we recall from Chapter 1 that the free energy is the component of the total energy available for performing work at isothermal conditions, (3.30) is an intuitively satisfying relation for a living system. It tells how fast the free energy is consumed in the system.

If (3.30) is applied to a subsystem such as the heart, it provides no more additional information than (3.17). Both equations describe, in slightly different ways, how the available stored chemical energy is consumed—to perform work and to overcome inefficiency or irreversibilities. The reader is urged to develop a firm understanding of these two relationships.

Returning to a description of the closed system energy balance for the human body, (3.10) and (3.13b), we recall that these equations describe the relation between basal metabolism, body temperature, heat loss, and work output. For long periods, the integrated forms of these equations may be used to estimate how much stored chemical energy is used. Rewriting (3.10) in its original form,

$$dU_c/dt + m_{\text{tot}}\, \hat{C}_p\, dT/dt = -\dot{Q}' - \dot{W} \tag{3.31}$$

Multiplying through by dt and integrating over some time period between meals of $t_2 - t_1$,

$$\int_{t_1}^{t_2} dU_c + m_{\text{tot}}\, \hat{C}_p \int_{t_1}^{t_2} dT = -\int_{t_1}^{t_2} \dot{Q}'\, dt - \int_{t_1}^{t_2} \dot{W}\, dt \tag{3.32}$$

which becomes

$$M_2 - M_1 + m_{\text{tot}}\, \hat{C}_p (T_2 - T_1) = -\int_{t_1}^{t_2} \dot{Q}'\, dt - \int_{t_1}^{t_2} \dot{W}\, dt \tag{3.33}$$

or

change in stored chemical energy	+	change in stored thermal energy	=	total heat lost to environment	+	total work performed

This relation can be used to explain quantitatively why fever and chills usually accompany each other in some succession during malfunctions in the physiology of the system. It is also of use in evaluating dietary requirements for different working and environmental situations.

Specific treatment of the actual computation of the heat loss term will be presented in the following chapters. It is only necessary to note here that the heat loss term in this closed system formulation has to account not only for heat transferred from the surfaces but also for the energy lost in vaporization of water in the respiratory system (recall Example 1.7), for the energy required to heat (or cool) the respiratory gases, and also for the energy lost in vaporization of water from the skin surfaces.

One final interesting example concerning the variation of basal metabolic energy conversion rate with body mass will be examined.

Example 3.5. Variation of Metabolic Rate with Body Mass

Figure 3.3 illustrates the observed relation between metabolic energy conversion rate and body mass for several different species, along with the variation of the metabolic rate per unit mass with body mass. It is noted that there is a logarithmic variation of both the total metabolic rate and the specific metabolic rate with total mass, as determined by measuring the total oxygen consumption.

A heuristic explanation of this relationship may be offered. The question is, How should one expect the metabolic rate to depend on body mass? In mathematical terms, How should $d\dot{M}_o/dm_{tot}$ depend

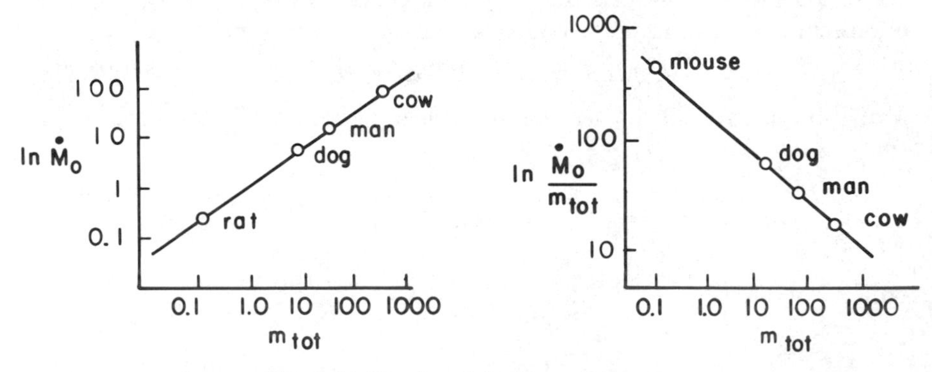

Fig. 3.3. Basal metabolism relations.

on m_{tot}? If one hypothesizes that this derivative should be proportional to metabolic rate, that is,

$$dM_o/dm_{tot} \propto \dot{M}_o/m_{tot} \tag{3.34}$$

on the grounds that larger metabolic rates may produce a larger rate of change of rate with mass while larger masses should produce smaller rates of change, then this relation can be integrated directly to produce

$$\ln \dot{M}_o = a \ln m_{tot} \tag{3.35}$$

which agrees with Figure 3.3 with $a = 0.75$. Further, the relation between $d\hat{\dot{M}}_o/dm_{tot}$, that is, the specific rate change with mass, can be directly worked out, the result being

$$\ln \hat{\dot{M}}_o = (a - 1) \ln m_{tot} = -0.25 \ln m_{tot} \tag{3.36}$$

which also agrees with Figure 3.3. The result, of course, is that the metabolic rate becomes proportional to mass to the 3/4 power, which has led many workers to use surface area as a correlative variable for metabolic rates, since for many geometries the total surface area is proportional to the mass to the 2/3 power. Although there is no strong theoretical basis for this correlation involving surface area, its use has been quite standard. The preceding analysis perhaps puts the consideration of rate variation on a stronger basis.

SUMMARY

In this chapter we have examined the transformation of energy in closed biological systems, that is, closed for the purposes of energy considerations. This chapter could have followed Chapter 1 directly since the discussion on performing material balances was not widely used in this treatment. However, it was thought that the energy analogies with the succeeding chapter were such that the presentation of this material here was appropriate.

While specifically applied to living systems, the philosophy of this chapter and the techniques of analysis offered could be similarly applied to any closed chemical system where energy was being transformed.

Bibliography

Background material on the energy relationships in man from the life scientist's point of view are available in many works. Two of the more quantitative treatments may be found in:

Guyton, A. C. *Textbook of Medical Physiology*. Saunders, 1966.

Ruch, T. C., and Patton, H. D. *Physiology and Biophysics*, 19th ed. Saunders, 1965.

In addition, general physiology textbooks such as the following can give important background on the mechanisms for energy storage and conversion:

Davson, H. *Textbook of General Physiology*. Little, Brown, 1964.
Dowben, R. J. *General Physiology*. McGraw-Hill, 1970.

chapter 4

Energy Balances in Open Systems

INTRODUCTION

In the previous chapter the first law of thermodynamics for a closed system was applied to a living system. The physical interpretations of this application were discussed, and it was pointed out that the modeling of a living system as closed is a serious approximation which in one sense is a contradiction in terms. Nevertheless, some insight into the transformation of stored chemical energy to do work and to operate a living system was gained, within the confines of the closed-system approximations. The concept that all the converted energy must eventually cross the system boundaries as heat or work and the basic understanding of the significance of the basal metabolism were perhaps the two major points in Chapter 3.

In this chapter, some general relations for energy conservation in open systems will be presented, and the examples of the previous chapter reexamined more thoroughly.

OPEN-SYSTEM ENERGY BALANCES

Before we reconsider living systems, it is necessary to extend the first law, as discussed in Chapter 1, to open systems. Of the several possible forms that can be written, we shall concentrate on two. The first is general for any system and may be written as

$$\frac{d}{dt}(U + \phi + K) = \sum_{\text{in}} (\hat{H} + \hat{\phi} + \hat{K})\, w - \sum_{\text{out}} (\hat{H} + \hat{\phi} + \hat{K})\, w + \dot{Q} - \dot{W} \quad (4.1)$$

or

$$\begin{matrix}\text{accumulation of} \\ \text{energy in the system}\end{matrix} = \begin{matrix}\text{total energy} \\ \text{carried in with mass in}\end{matrix} - \begin{matrix}\text{total energy} \\ \text{carried out}\end{matrix}$$

$$+ \text{heat in} - \text{work out}$$

A second form of this overall energy balance that will be of some use is a special form for isothermal isobaric systems.

$$\frac{d}{dt}(G + \phi + K) = \sum_{\text{in}} (\hat{G} + \hat{\phi} + \hat{K})\, w - \sum_{\text{out}} (\hat{G} + \hat{\phi} + \hat{K})\, w - \dot{W} - \dot{R} \quad (4.2)$$

Note that (4.1) is written in terms of internal energy and enthalpy and contains a term for heat losses; whereas (4.2), the isothermal equivalent of (4.1), is written in terms of free energy and instead of heat losses contains a term for the irreversibilities in the system. This point will be expanded later.

The steady state form of (4.1) is an important relationship for many engineering applications. When the total energy content of the system is not changing with time, (4.1) may be rewritten as

$$\sum_{\text{out}} (\hat{H} + \hat{\phi} + \hat{K})\, w = \sum_{\text{in}} (\hat{H} + \hat{\phi} + \hat{K})\, w + \dot{Q} - \dot{W} \quad (4.3)$$

This expression relates the total energy entering and leaving the system with the heat transferred and work done. For cases where no external work is performed (such as in chemical reactors and heat exchangers), and the mechanical energy of the input and output streams are negligible, (4.3) may be written

$$\sum_{\text{out}} \hat{H}\, w - \sum_{\text{in}} \hat{H}\, w = \dot{Q} \quad (4.4)$$

Because only the enthalpy appears on the left, this relation is often incorrectly referred to as an enthalpy balance. It is instead merely a special form of a total energy balance.

One important point which should be made here is that (4.4) is completely adequate to handle systems in which chemical reactions occur, without adding any terms to account for the "heat generated by chemical reaction." The enthalpy change upon reaction can be treated correctly by the correct evaluation of the enthalpy of the streams that enter and leave the system, as will be illustrated in one of the following examples.

Some examples of the use of these first four energy balance relations for open systems will now be presented.

Example 4.1. Blood Cooler

In this example we shall compute the heat output required to cool 500 ml/min of blood from 37°C to 30°C in an extracorporeal heat exchanger.

The appropriate form to use is (4.4), since this is a steady state problem, with one input and one output stream with negligible mechanical energy changes. Since the density of blood is about 1 gm/ml, $w_{in} = w_{out} = 500$ gm/min. Then, (4.4) may be rewritten

$$\dot{Q} = w\, \hat{C}_p\, (T_{out} - T_{in})$$

$$\dot{Q} = 500 \text{ gm/min } (1.0 \text{ cal/gm } ^\circ\text{C})(-7^\circ\text{C})$$

$$\dot{Q} = -3500 \text{ cal/min} = -3.5 \text{ kcal/min} \tag{4.5}$$

which says that 3.5 kcal of energy must be removed as heat every minute in order to cool the blood the desired amount. This is perhaps the simplest of all possible applications of the overall energy balance equations, but the principle as illustrated here can be extended to any process with any number of input and output streams.

Example 4.2. Cooling of a Stirred Tank

In this example we shall compute the rate at which an insulated well-stirred tank will be cooled if the input stream of 10 lb/min comes in at 60°F and the output stream leaves at essentially the tank temperature. The tank contains 100 lb of water and is initially at 100°F.

This is an unsteady-state problem. The starting point is (4.1), and the appropriate form is

$$dU/dt = \sum_{in} \hat{H}\, w - \sum_{out} \hat{H}\, w \tag{4.6}$$

The total internal energy of the system will be the product of its mass, its heat capacity, and its temperature relative to some reference temperature. In this example, the choice of the reference temperature is immaterial since, as the reader can confirm, it drops out from both sides of (4.6). Assuming that the temperature of the tank's contents and its output stream can be described by a single temperature T and noting that the mass in the tank at any time is constant, then (4.6) may be written

$$m\, \hat{C}_p\, dT/dt = w\, \hat{C}_p\, (T_{in} - T) \tag{4.7}$$

and substituting the values given above,

$$dT/dt = 0.10\,(60 - T) \tag{4.8}$$

where m/w = 10 min = the time constant for this process. This result shows that the cooling rate for the tank is proportional to time and that it decreases from an initial value of -4° F/min to zero after a very long time. The solution of this differential equation, with its initial condition $T = 100$ at $t = 0$, is $T = 60 + 40 \exp(-0.1\, t)$.

Example 4.3. Adiabatic Flame Temperature

A common problem in chemical engineering is the calculation of the adiabatic flame temperature of the chemical reaction taking place in a combustor or other reacting system. The basic relationship used to perform this calculation is (4.4), with $\dot{Q} = 0$. This must be solved along with the material balance relations that apply in order to estimate the maximum temperature in the system. A typical example follows.

Suppose one mole of methane (CH_4) is completely burned in air in a combustor using a stoichiometric ratio of oxygen to fuel. What will be the maximum temperature that will exist in the combustor? How much heat must be removed if the products of the combustion are cooled to the same temperature as the reactants?

The starting point is the writing of the overall chemical reaction:

$$CH_4\,(g) + 2\,O_2\,(g) \longrightarrow CO_2\,(g) + 2\,H_2O(g) \qquad (4.9)$$

Note that the reaction must be balanced and the states indicated. It will be most convenient to work this problem in molal units, rather than mass and to take for a basis one mole of methane. Since air is being used, it will be necessary to account for the nitrogen that "rides along" in the combustor. The moles of nitrogen that will be involved may be quickly computed as

$$2.0 \text{ moles } O_2 \times 79/21 \text{ moles } N_2 \text{ /mole } O_2 = 7.52 \text{ moles } N_2$$

The maximum temperature will of course result if the reaction goes 100% to completion and $\dot{Q} = 0$. Then the enthalpy of the products will equal the enthalpy of the reactants plus the enthalpy change on reaction, that is, using one mole of CH_4 as the basis:

$$H_{out} = H_{in} = \sum_{in} \widetilde{H}_i n_i \,(@\, T_{in}) - (\Delta \widetilde{H}_R) = \sum_{out} \widetilde{H}_i n_i \,(@\, T_{out}) \qquad (4.10)$$

which comes from the definition of the enthalpy change on reaction and the consideration that enthalpy is a point or state function. Another way of interpreting this is to think of the enthalpy change of reaction as being used adiabatically to raise the products up to the adiabatic flame temperature.

If the reactants are brought in at 25°C, then the problem is somewhat simplified. The enthalpy of the products will then be equal to the enthalpy change of reaction which delivers gaseous products at 25°C. The standard heat of combustion, (introduced in Chapter 1) for methane is -212.8 kcal/mole of methane based on gaseous CO_2 and liquid water as the products. So, this value must be corrected for the enthalpy required to vaporize the water. The calculation then proceeds as follows:

Moles N_2 out = 7.52 Mean molal heat capacity = 7.99 cal/mole °K
Moles H_2O out = 2.0 Mean molal heat capacity = 10.43
Moles CO_2 out = 1.0 Mean molal heat capacity = 13.10

Molal standard heat of combustion of methane = -212.8 kcal/mole
Molal heat of vaporization of water = 9.717 kcal/mole at 100°C
Mean molal heat capacity of liquid water = 18 cal/mole °K

Substituting into (4.10), correcting the heat of reaction, and representing the final temperature difference as $T - 25$,

$$-(-212{,}800) - 2.0(9{,}717) - 2(18)(75) = 7.99(7.52)(T - 25) + 13.10(1.00)(T - 25) + 10.43(2.00)(T - 25)$$

$$T - 25 = 190{,}666/94.06 = 2{,}020 \text{ °C}$$

Adiabatic flame temperature:

$$T = 2{,}045 \text{ °C}$$

It should be noted that:

1. If the combustion took place with pure oxygen instead of air, the resultant adiabatic temperature would be higher, since enthalpy would not be used up to heat the nitrogen.
2. The mean heat capacities used are the average value of the capacities between 25 and 2,000 °C.
3. In the actual case the reaction does not go 100% to completion, and correcting for this gives a value of T = 1,918°C.

LIVING SYSTEMS

The next task is to apply these overall energy balances for open systems to living systems and to extend our limited treatment of these systems that we began in Chapter 3. As indicated previously, in almost every case changes in the kinetic and potential energy of living systems can be neglected. In addition, the potential and kinetic energies associated with the material that enters and leaves the system are usually very small for living systems. It should be pointed out once again that this is not necessarily the case with many engineering systems—in fact, changes in the mechanical energies of the system and the magnitudes of such energies associated with the streams entering and leaving the system in items like turbines, pumps, ejectors, and propulsion devices are usually the most important terms in the overall energy balance. For most biological systems however, (4.1) reduces to

$$dU/dt = \sum_{\text{in}} \hat{H}\, w - \sum_{\text{out}} \hat{H}\, w + \dot{Q} - \dot{W} \tag{4.11}$$

Furthermore, since most biological systems operate isothermally and isobarically and at constant volume, changes in enthalpy and internal energy are equivalent. Further, we can recall from Chapter 3 the substitutions for dU/dt and $\dot{Q}$ and rewrite (4.11) as

$$\dot{M}_o + \Delta\dot{M} + \dot{T} = \sum_{\text{in}} \hat{U}\, w - \sum_{\text{out}} \hat{U}\, w - \dot{Q}' - \dot{W} \tag{4.12}$$

The next task is to sum up the input and output streams which carry energy into and out of the system. These streams have already been discussed in Example 2.3 in relation to the overall water balance. We will summarize them here and discuss their energy content.

Input Streams

1. *Food and drink*, denoted by w_f. Certainly this is the major energy input to living systems. Unfortunately, the concept of a flow rate w_f is uncomfortable when referring to food, and the integrated form of (4.12) will be more appropriate for consideration of this term.
2. *Inspired air*, denoted by w_I. The energy content of this stream is usually less than that of the expired gases, but it is necessary to account for it in the overall balance.
3. *Injections and transfusions* will be accounted for in the food and drink term.

Output Streams

1. *Expired respiratory gases*, denoted by w_E. As demonstrated in Example 1.7, the net energy loss due to the difference in energy content between this stream and the inspired gases can be appreciable.

2. *Wastes*, denoted by w_w. In this category we include feces, urine, and sweat runoff. If our reference temperature for computing internal energy content is taken as body temperature, then the energy content of this stream will be very small and, except for the combustion value of the solid material in the feces, is usually zero.
3. *Evaporation through the skin*, denoted by w_v. This stream, which accounts normally for about 10% of the daily transfer to the environment for humans, is often thought of as a heat loss term but is more fundamentally a stream of material leaving the system. The energy loss by this mechanism will be the product of the enthalpy of vaporization of water at the operating temperature and the mass rate at which water is evaporated.

It is convenient to treat the energy from the food stream as an enthalpy flow rate contribution and to consider the other streams as

$$\hat{H}_I\, w_I - \hat{H}_E\, w_E = -\dot{Q}_{\text{resp}} \qquad (4.13)$$

$$-\hat{H}_v\, w_v = -\dot{Q}_{\text{vap}} \qquad (4.14)$$

$$-\hat{H}_w\, w_w = 0 \qquad (4.15)$$

Then (4.12) becomes

$$\dot{M}_o + \Delta\dot{M} + \dot{T} = \hat{H}_f\, w_f - \dot{Q}_{\text{resp}} - \dot{Q}_{\text{vap}} - \dot{Q}' - \dot{W} \qquad (4.16)$$

The first three terms on the right can now be expressed in more exact terms. The energy content of food can be expressed by recalling the results of Example 1.6:

$$\hat{H}_f = \hat{U}_f = 4.1\, x_p + 4.1\, x_c + 9.3\, x_f \text{ kcal/gm} \qquad (4.17)$$

where x_p, x_c, and x_f represent the weight fractions of protein, carbohydrate, and fat respectively of the ingested food.

The respiratory energy loss term can be written, recalling the results of Example 1.7, as

$$\dot{Q}_{\text{resp}} = \dot{V}/\widetilde{V}\, [\widetilde{C}_p\, (T_E - T_I) + (Y_E - Y_I)\, \Delta\widetilde{H}_v] \qquad (4.18)$$

where

$\dot{V}$ = ventilation rate, liters/min
$\widetilde{V}$ = specific volume of inspired gas, liters/mole
$\widetilde{C}_p$ = molal heat capacity of inspired gas, kcal/mole °C
T_E = temperature of expired gas, °C
T_I = temperature of inspired gas, °C
Y_E = mole ratio of H_2O in expired gas, moles H_2O/mole dry gas
Y_I = mole ratio of H_2O in inspired gas, moles H_2O/mole dry gas
ΔH_v = molal heat of vaporization of H_2O, kcal/mole

For humans breathing air,

$$\dot{Q}_{\text{resp}} = \dot{V}/\tilde{V}\,[0.007(37 - T_I) + 10.8(0.066 - Y_I)]\ \text{kcal/min} \qquad (4.19)$$

The reader should check the results of Example 1.7 using (4.19).

Thorough consideration of the vaporization term $\dot{Q}_{\text{vap}}$ will have to be postponed until Chapter 7. The rate at which vaporization takes place will depend not only on how much area is exposed and on other properties of the system but also on the environmental factors that influence the rate of mass transfer away from the system.

As in the closed-system consideration of energy in living systems, an integrated form of (4.16) is a most useful relation. Proceeding as before, we obtain

$$M_2 - M_1 + \overline{T}_2 - \overline{T}_1 = \hat{H}_f m_f - \int_{t_1}^{t_2} (\dot{Q}_{\text{resp}} + \dot{Q}_{\text{vap}} + \dot{Q}' + \dot{W})\,dt \qquad (4.20)$$

Since we are considering open systems, it is necessary to accompany this energy balance with a material balance. The following form of (2.1) is appropriate here:

$$dm_{\text{tot}}/dt = \sum_{\text{in}} w - \sum_{\text{out}} w \qquad (4.21)$$

with its associated integrated form,

$$\Delta m_{\text{tot}} = \sum_{\text{in}} m - \sum_{\text{out}} m \qquad (4.22)$$

The symbol $\overline{T}$ in (4.20) denotes the total thermal energy, $m_{\text{tot}}\,\hat{C}_p T$. The input and output streams needed in (4.21) have previously been discussed in Example 2.3 and in the derivation of (4.16).

Equations (4.20) and (4.22), along with (4.17) and (4.19), are the basic ones that relate diet, exercise, body weight, and environment. Three additional pieces of information are required to completely close the description. The first, to be discussed extensively in the following two chapters, deals with the computation of the amount of energy lost as heat to the environment, and the ability to predict this quantity for wide ranges of environmental conditions. The second, a much more difficult problem, deals with the definition and computation of the work term. The physical definition of net work is inadequate for use in the overall energy balance for living systems. A man climbing stairs and returning has done no net work according to an overall description based solely on the usual definition of external work, yet his muscle system has experienced considerable net work output. The relationship between physiological work and external activity is a subject of intense research activity.

The third piece of information, the implications of which are concealed in (4.20) and (4.21), deals with exactly how the system chooses to adjust priorities with regard to adjusting chemical composition to provide energy for its various activities. Interrelated with this is the concept of the efficiency of the various muscle systems of the body. Apparently, there is no a priori way to predict these priorities of efficiencies for a given individual, and this in part is critical in the explanation of the great deviations between individuals as to their weight-diet-exercise characteristics.

Despite these limitations, these equations form the basis for an understanding of the mass-energy relations in living systems and for continued research in this area.

Further comment on the terms on the left side of (4.20) is required. Changes in the quantities $\overline{M}$, $(\overline{M}_2 - \overline{M}_1)$ and $\overline{T}$, $(\overline{T}_2 - \overline{T}_1)$ can reflect changes in mass as well as changes in composition and temperature. Recalling from Chapter 3,

$$U = \hat{U}\, m_{\text{tot}} = \hat{U}_c\, m_{\text{tot}} + \hat{U}_T\, m_{\text{tot}} \tag{4.23}$$

so that

$$\Delta U = \Delta (\hat{U}\, m_{\text{tot}}) = \Delta (\hat{U}_c\, m_{\text{tot}}) + \Delta (m_{\text{tot}}\, \hat{C}_p\, T) \tag{4.24}$$

that is,

$$U_2 - U_1 = \hat{U}_{c_2}\, m_{\text{tot}_2} - \hat{U}_{c_1}\, m_{\text{tot}_1} + \hat{C}_p\, m_{\text{tot}_2}\, T_2 - \hat{C}_p\, m_{\text{tot}_1}\, T_1 \tag{4.25}$$

The point is that even if the temperature and chemical composition remain constant, changes in the total mass of the system change the amount of stored chemical and thermal energy. This is important in understanding how the intake of food adds to the stored chemical energy of the system, not only by addition of high energy compounds but also by the addition of mass.

In the previous chapters as well as in this one we have referred to "normal humans" and have used typical numerical values for certain physiological parameters in the example problems. It is useful to assemble and present here a consistent set of these values and refer to a *standard man*. Such a set has been assembled from the physiological literature and is presented in Table 4.1.

It is also useful to summarize the important energy conservation relations and the associated equations from the last two chapters. The general energy balance equations are presented in Table 4.2 and the special forms for living systems in Table 4.3.

Table 4.1. Standard Man Data

Age	30 yr
Height	5 ft 8 in.
Weight	150 lb_m
Surface area	1.80 m^2
Normal body core temperature	37.0°C
Normal mean skin temperature	34.2°C
Heat capacity	0.86 kcal/kgm °C
Percent body fat	15%
Basal metabolism	40 kcal/m^2hr, 72 kcal/hr, 1730 kcal/day
Oxygen consumption	250 ml/min (@ STP)
CO_2 production	200 ml/min
Respiratory quotient	0.80
Blood volume	5 liters
Resting cardiac output	5 liters/min
Systemic blood pressure	120/80 mm Hg
Heart rate	65/min
Total lung capacity	6,000 ml
Vital capacity	4,200 ml
Ventilation rate	6,000 ml/min
Alveolar ventilation rate	4,200 ml/min
Tidal volume	500 ml
Dead space	150 ml
Breathing frequency	12/min
Pulmonary capillary blood volume	75 ml
Arterial O_2 content	0.195 ml O_2/ml blood
Arterial CO_2 content	0.480 ml CO_2/ml blood
Venous O_2 content	0.145 ml O_2/ml blood
Venous CO_2 content	0.520 ml CO_2/ml blood

Table 4.2. Summary of Energy Balances

Closed systems:

$$d/dt\ (U + \phi + K) = \dot{Q} - \dot{W}$$

Open systems:

$$d/dt\ (U + \phi + K) = \sum_{\text{in}} (\hat{H} + \hat{\phi} + \hat{K})\, w - \sum_{\text{out}} (\hat{H} + \hat{\phi} + \hat{K})\, w + \dot{Q} - \dot{W}$$

Isothermal:

$$d/dt\ (G + \phi + K) = \sum_{\text{in}} (\hat{G} + \hat{\phi} + \hat{K})\, w - \sum_{\text{out}} (\hat{G} + \hat{\phi} + \hat{K})\, w - \dot{W} - \dot{R}$$

Table 4.3. Summary of Material and Energy Balances for People

Closed-system approximation:

$$\dot{M}_o + \Delta\dot{M} + \dot{T} = -\dot{Q}' - \dot{W}'$$
$$\dot{M}_o = -\dot{Q}'_o = -72 \text{ kcal/hr}$$
$$\dot{T} = m_{\text{tot}}\, \hat{C}_p\, dT/dt$$
$$\dot{M} = \dot{M}_o + \Delta\dot{M} = dU_c/dt$$
$$-\dot{W} = b\,\Delta\dot{M} \quad b < 1$$
$$\dot{G} = -\dot{W} - \dot{R}$$

Integrated:

$$M_2 - M_1 + m_{\text{tot}}\, \hat{C}_p\, (T_2 - T_1) = -\int_{t_1}^{t_2} (\dot{Q}' + \dot{W})\, dt$$

Open-system treatment:

$$dm_{\text{tot}}/dt = \sum_{\text{in}} w - \sum_{\text{out}} w$$

$$dm_i/dt = \sum_{\text{in}} w\,\omega_i - \sum_{\text{out}} w\,\omega_i + \dot{r}_i$$

$$\dot{M}_o + \Delta\dot{M} + \dot{T} = \hat{H}_f\, w_f - \dot{Q}_{\text{resp}} - \dot{Q}_{\text{vap}} - \dot{Q}' - \dot{W}$$
$$\hat{H}_f = 4.1\, x_p + 4.1\, x_c + 9.3\, x_f$$
$$\dot{Q}_{\text{resp}} = \dot{V}/\widetilde{V}\,[0.007(37 - T_I) + 10.8(0.066 - Y_I)]$$

$$M_2 - M_1 + \overline{T}_2 - \overline{T}_1 = \hat{H}_f\, m_f - \int_{t_1}^{t_2} (\dot{Q}_{\text{resp}} + \dot{Q}_{\text{vap}} + \dot{Q}' + \dot{W})\, dt$$

SUMMARY

This chapter completes our treatment of material and energy balances. Now we must turn our attention to rate processes, that is, an analysis of the fundamental physical processes of heat and mass transfer. The first four chapters have been necessary to lay a foundation and to illustrate the context for the study of rate processes to follow.

Bibliography

An excellent text which summarizes energy balances in open systems in a complete and concise format is:

Bird, R. B., Stewart, W. E., and Lightfoot, E. N. *Transport Phenomena.* Wiley, 1960.

Another valuable source for information on the type of techniques discussed in this chapter is:

Hougen, O. A., Watson, K. M., and Ragatz, R. A. *Chemical Process Principles*. Wiley, 1943.

Data for the "standard man" were extracted from:

Comroe, J. H., Jr. *Physiology of Respiration*. Year Book Medical Publishers, 1965.

Guyton, A. C. *Textbook of Medical Physiology*. Saunders, 1966.

chapter 5

Fundamentals of Heat Transfer

INTRODUCTION

The importance of heat transfer in living systems has been demonstrated. In addition to living systems, the consideration of heat transfer in the design and operation of artificial organs, life support systems, and other biomedical equipment is often a vital part of the success and the optimization of the system. In this chapter we will present the basic principles involved in the transfer of thermal energy. Chapter 6 will cover physiological and biomedical applications and other uses of these principles.

In Chapter 1 heat was defined as energy transfer resulting from a difference in temperature between the system and the surroundings. It was also pointed out that the term heat transfer was literally redundant. Strictly speaking, energy transfer due to vaporization from system surfaces is not heat transfer, although in Chapters 3 and 4 we found it convenient to consider it as such. In this chapter we will restrict ourselves to "legitimate" heat transfer and discuss only temperature-driven energy flows. There are three modes by which energy may be transferred either within or between systems as the result of temperature gradients or differences.

Conductive heat transfer, or *conduction*, refers to the transport of energy due to molecular collisions within the material experiencing a temperature gradient. This transport process is strictly molecular, and the properties of the material in conjunction with the temperature field are sufficient to describe the process.

Convective heat transfer, or *convection*, is the transport of energy brought about by the hydrodynamic or bulk motion of the fluid media (liquid or gas) that is responsible for the heat flow. We differentiate between *forced convection*, in which the fluid motion is caused by an imposed pressure gradient, and *free convection*, in which the motion is caused by buoyancy forces caused by density differences in the fluid.

Radiative heat transfer, or *radiation*, refers to the transport of energy as the result of electromagnetic radiation (the transfer of photons). It differs mainly from the other two modes in that it requires no intermediate material phase between the radiating or exchanging surfaces.

All three of these modes are very important in living systems and also in virtually every system in the physical world. An understanding of the details of each mode and the differences between them is crucial to the design and operation of engineering systems involving heat transfer as well as to the understanding of energy relationships in living systems.

CONDUCTIVE HEAT TRANSFER

To understand the mechanism of conductive heat transfer, it is necessary to think of temperature as a measure of molecular activity; that is, an increase in the temperature of a substance represents an increase in the kinetic energy and the activity of the molecules of the substance. With this picture in mind, we can understand that the transfer of energy takes place in a material as the result of collisions between higher energy molecules and ones of lower energy and that we can say that "heat" is transferred only in the direction of decreasing temperatures. The net flow of energy in this situation will be downhill, that is, down the temperature gradient.

The basic relationship between the temperature gradient in a material and the rate of heat flow was experimentally determined by Fourier, who observed that for a given material the rate of energy transfer was directly proportional to the temperature difference across the material and inversely proportional to the thickness of the material. Fourier called the proportionality constant the "conducibility" of the material. This name has evolved to the term *thermal conductivity*, and Fourier's law, which defines it, may be written in one-dimensional differential form as

$$q_x = \dot{Q}/A = -k\ (dT/dx) \tag{5.1}$$

where

q_x = heat flux, kcal/m^2 min in the x direction
$\dot{Q}$ = heat flow, kcal/min
A = normal area through which the heat is transferred, m^2
k = thermal conductivity, kcal/m min °C
dT = differential of temperature, °C
dx = thickness differential, m

The negative sign is used so that the heat flux will be positive in the direction of the negative temperature gradient, that is, the energy flows downhill.

If (5.1) is integrated over a thickness x, the edges of which are held at T_o and T_L and which has a constant value of thermal conductivity k, the result is

$$q_x = k\left(\frac{T_o - T_L}{x}\right) \tag{5.2}$$

which is actually the original form of Fourier's law.

Equation 5.2 is the first of several rate equations which are fundamental to the study of the transport of energy and material. The general form of such expressions is

$$\text{Rate of transfer } \alpha \ \frac{\text{driving force}}{\text{resistance}}$$

or more familiarly,

$$\text{Rate of transfer } \alpha \text{ conductance} \times \text{driving force}$$

It is quite evident which of these quantities are which in (5.2). The *rate* is the rate at which heat is transferred, the *conductance* is the thermal conductivity per unit length, and the *driving force* is the temperature difference. Ohm's law for the conduction of electricity is another and perhaps more familiar form of a rate expression.

Some selected values of the thermal conductivity k are presented in Table 5.1. In general, thermal conductivities of solids are about one hundred times those of liquids, which are in turn about one hundred times those of gases. The thermal conductivity of a sub-

Table 5.1. Selected Values of Thermal Conductivity

Substance	*Temperature*	*Thermal Conductivity k*
Air	300°K	4.25×10^{-5} cal/cm sec °C
O_2	200°K	4.38
O_2	300°K	6.35
CO_2	300	3.98
Water	100°C	1.60×10^{-3}
Water	20	1.43
Benzene	20	3.78
Copper	20	0.918×10^{0}
Graphite	20	0.012
Steel	20	0.112
Steel	100	0.107
Biological tissue	37°C	1.10×10^{-3} cal/cm sec °C 0.40 kcal/m hr °C

stance generally increases with increases in temperature, although for many solids, particularly insulating materials, it decreases with temperature.

In *steady-state conductive* heat transfer situations, where the temperature profile is not a function of time, and in the absence of any internal heat generation, the heat flow remains constant. That is, the input and output of heat through a given slice of material are equal. The following set of conditions then results:

Rectangular geometry:

$$q_x = \text{constant} \tag{5.3a}$$

Cylindrical geometry:

$$r\, q_r = \text{constant} \tag{5.3b}$$

Spherical geometry:

$$r^2\, q_r = \text{constant} \tag{5.3c}$$

In Equations 5.3a–c, while the total *heat flow* is constant, the *flux*, or the flow per unit area, must vary in curvilinear coordinates due to the increasing area as the radial dimension is increased.

For situations where the thermal conductivity may be assumed constant, the following useful set of flux expressions and temperature profiles are then implied by 5.3a–c:

Rectangular geometry ($T = T_o$ at $x = x_o$):

$$q_o = -k\,(dT/dx)_{x=x_o}$$

$$T - T_o = \frac{-q_o}{k}\,(x - x_o) \tag{5.4a}$$

Cylindrical geometry ($T = T_o$ at $r = r_o$):

$$q_o = -k(dT/dr)_{r=r_o}$$

$$T - T_o = \frac{-q_o r_o}{k}\,\ln r/r_o \tag{5.4b}$$

Spherical geometry ($T = T_o$ at $r = r_o$):

$$q_o = -k\,(dT/dr)_{r=r_o}$$

$$T - T_o = \frac{-q_o r_o^2}{k}\,(1/r - 1/r_o) \tag{5.4c}$$

These relations, which describe the shape of the steady-state temperature profiles for these geometries, can also be "turned inside out" and solved directly for the heat flow at the boundary, designated by the subscript $_o$, if the temperature at the other boundary, $x = x_1$ or $r = r_1$, is specified as T_1. Then:

Rectangular geometry:

$$q_o = -k\,(T_1 - T_o)/(x_1 - x_o) \tag{5.5a}$$

Cylindrical geometry:

$$q_o = -\frac{k}{r_o}\,(T_1 - T_o)/\ln r_1/r_o \tag{5.5b}$$

Spherical geometry:

$$q_o = -\frac{k}{r_o^2}\,(T_1 - T_o)/(1/r_1 - 1/r_o) \tag{5.5c}$$

These six basic one-dimensional equations may be easily extended to include composite materials, since at steady state the heat flow through each layer will be the same.

Example 5.1. Heat Transfer through Composite Plane Walls

Figure 5.1 illustrates a heat transfer problem which occurs fre-

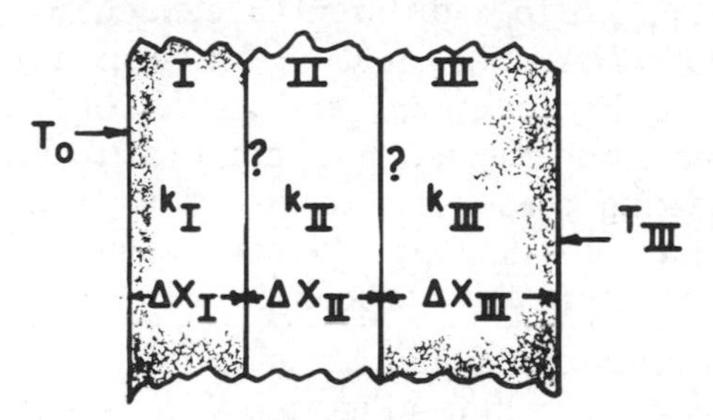

Fig. 5.1. Composite wall heat conduction.

quently in many systems. The temperatures at the inside and outside are known, but the intermediate temperatures at the interfaces between the layers of different material are not. Knowing the relative dimensions and properties of the layers, it is desired to compute the heat flow through the walls. This may be accomplished by writing Equation 5.4a for each layer.

$$T_o - T_I = q_o\,\Delta x_I/k_I \tag{5.6a}$$

$$T_I - T_{II} = q_o\,\Delta x_{II}/k_{II} \tag{5.6b}$$

$$T_{II} - T_{III} = q_o\,\Delta x_{III}/k_{III} \tag{5.6c}$$

Since

$$T_o - T_{III} = (T_o - T_I) + (T_I - T_{II}) + (T_{II} - T_{III}) \tag{5.7}$$

we can substitute and rearrange, equating the fluxes in each section, and obtain

$$q_o = \frac{T_o - T_{III}}{\Delta x_I/k_I + \Delta x_{II}/k_{II} + \Delta x_{III}/k_{III}} \tag{5.8}$$

The denominator of this result can be seen to be a sum of the three resistances in the problem. The reader will probably notice the direct analogy between this result and the flow of current through a series of three resistances for a given total voltage drop. The reader should work out the details of this analogy for himself.

When the temperature is also a function of time as well as position in a material, the consideration of the heat flow–temperature problem becomes more complex. The analysis of such situations is based on the principle

$$\begin{matrix}\text{Accumulation} \\ \text{of heat}\end{matrix} = \begin{matrix}\text{rate of} \\ \text{heat in}\end{matrix} - \begin{matrix}\text{rate of} \\ \text{heat out}\end{matrix} + \begin{matrix}\text{rate of heat} \\ \text{generation}\end{matrix} \tag{5.9}$$

which may be applied to a differential element of the material to produce a partial differential equation which in turn must be solved along with the required boundary and initial conditions. This often requires relatively sophisticated mathematical techniques. A fairly common engineering practice is to resort to available generalized graphical solutions for such problems. The following example demonstrates the use of such a procedure.

Example 5.2. Conductive Cooling of a Sphere

In this example we will use Figure 5.2 to determine the transient temperature profile in a sphere initially at a uniform temperature $T_o = 100°$F and subjected to an external temperature $T_1 = 50°$C. We will consider the particular problem where the properties of the sphere are uniform and where the surface can be held at T_1 with no appreciable thermal resistance.

Figure 5.2 is the graphical dimensionless solution of the partial differential equation which describes the situation:

$$\rho \hat{C}_p \, (\partial T/\partial t) = (k/r^2) \, (\partial/\partial r) \left(r^2 \, \frac{\partial T}{\partial r} \right) \tag{5.10}$$

with the boundary conditions

$$\begin{aligned} T &= T_1 \quad \text{at } r = R, \ t = t \\ T &= T_o \quad \text{at } r = r, \ \ t = 0 \\ \partial T/\partial r &= 0 \quad \text{at } r = 0, \ t = t \end{aligned}$$

For a sphere of $R = 10$ cm, $k = 0.60$ cal/gm cm °K, $\rho = 5$ gm/cm^3, and $\hat{C}_p = 0.60$ cal/gm°C, let us compute the temperature at the

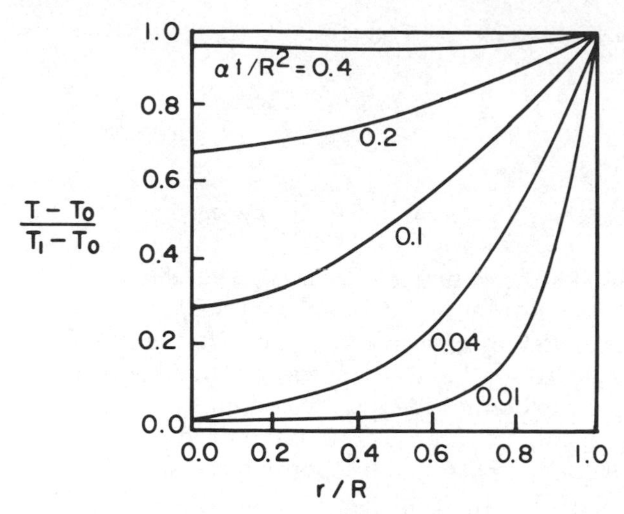

Fig. 5.2. Transient heat conduction solutions. T=temperature at t, r; T_o=temperature at t=0; T_1=temperature at r=R; R=sphere radius; α=$k/\rho C_p$; t=time.

center after 100 sec for the temperatures stated earlier. Then, $\alpha\, t/R^2 = 0.20(100)/100 = 0.20$, and from Figure 5.2 at $r/R = 0$, the temperature ratio is read as 0.70. Then

$$(T - T_o)/(T_1 - T_o) = 0.70 \quad T = -50(0.7) + 100 = 65°\text{F}$$

Steady-state heat conduction problems with internal generation may often be solved directly. These are of interest in many physiological applications since, as we have seen in the previous chapters, the metabolic energy conversion mechanism takes place within the body phase. Here we shall present some general results for heat generation problems and look at one example of interest.

Example 5.3. Temperature Rise in a Working Muscle

In this example we will estimate the temperature rise in a working muscle fiber, which we will model as a cylindrical rod. There will be an internal heat generation due to contracting fibers, which we will model as uniform. We will use this example to introduce the special forms of (5.3a–c) for the case where there is a uniform energy generation term.

Rectangular geometry:

$$-k\,\frac{d^2 T}{dx^2} = \dot{S} \tag{5.11a}$$

Cylindrical geometry:

$$-\frac{k}{r}\frac{d}{dr}\left(r\frac{dT}{dr}\right)=\dot{S} \tag{5.11b}$$

Spherical geometry:

$$\frac{k}{r^2}\frac{d}{dr}\left(r^2\frac{dT}{dr}\right)=\dot{S} \tag{5.11c}$$

$\dot{S}$ represents the volumetric rate of heat generation by electrical, chemical, or nuclear means (cal/cm³ sec).

For our model muscle we will assume that the temperature of the surface is held at $T = 37°\text{C}$ and compute the muscle temperature rise if the $\dot{S}$ due to metabolic energy conversion is 5 cal/cm³ hr, or 5 times the resting metabolic rate. The diameter will be taken as 2 cm, and the thermal conductivity as 0.001 cal/cm sec °C.

Integrating (5.11b),

$$-\int_0^r d\left(r\frac{dT}{dr}\right)=\int_0^r \frac{\dot{S}\,r}{k}\,dr \tag{5.12a}$$

which produces

$$-r\frac{dT}{dr}=\frac{\dot{S}\,r^2}{2k} \tag{5.12b}$$

Integrating a second time and making use of the surface boundary condition,

$$\int_{T_S}^{T} dT=\int_R^r -\frac{\dot{S}\,r}{2k}\,dr \tag{5.12c}$$

$$T-T_S=\frac{\dot{S}}{4\,k}(R^2-r^2) \tag{5.12d}$$

The maximum temperature at $r = 0$ will be

$$\begin{aligned} T &= T_s+\dot{S}\,R^2/4\,k = 37 + 5 \times 1/4 \times 0.001 \times 3{,}600 \\ &= 37 + 0.33 = 37.33°\text{C} \end{aligned} \tag{5.12e}$$

The maximum temperature rise for this condition is about one-third of a degree.

Note that (5.12b) can be used to obtain the heat flux at the muscle surface, since

$$q_{r=R}=-k(dT/dr)_{\text{at}\,r=R}=\dot{S}R/2$$

so that the total heat generated per unit length of muscle is either computed from flux × area = $\dot{S}R/2 \times 2\pi R = \pi R^2\dot{S}$, or by multiplying the volumetric rate $\dot{S}$ by the volume πR^2.

Equations 5.11a–c have general utility in the solution of one-dimensional steady-state conductive heat transfer problems with internal generation.

CONVECTIVE HEAT TRANSFER

The mechanism of convective heat transfer is the bulk hydrodynamic motion of the surrounding fluid medium near the surface from which heat is being transferred. We usually are concerned with transfer away from a surface—either inside or outside a conduit or container or away from an object, with the adjacent fluid being responsible for carrying the energy away (or bringing it to the surface). Although the transfer at the "local level" (that is, in the limit) is by conduction, the actual velocity of the surrounding fluid is usually the controlling variable.

Since the actual solution of the fundamental differential equations which describe convective heat transfer situations is extremely complicated, engineers and physicists, following the original suggestion of Newton, resort to the use of *heat transfer coefficients* for these problems. The concept of such a coefficient is shown in Figure 5.3, which depicts a temperature profile near a solid surface

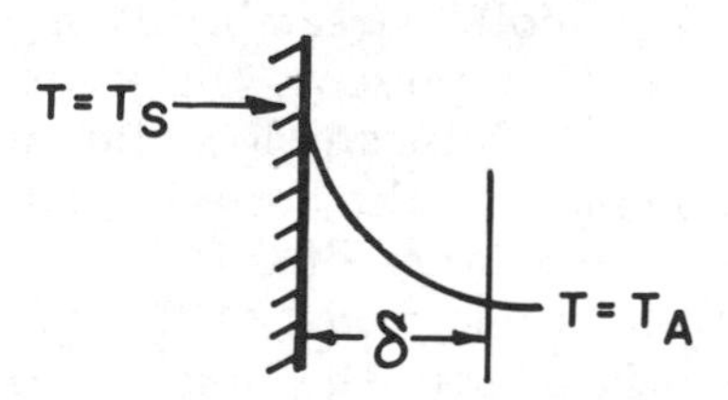

Fig. 5.3. Film model for heat transfer.

that is undergoing convective cooling. The heat flux away from the surface could theoretically be computed from Fourier's law by

$$q_x = k\,(T_S - T_A)/\delta \qquad (5.13)$$

The drawback is that usually the thickness δ, referred to as the film thickness, is not known and is difficult to measure or estimate for most situations. Therefore, as an alternative to (5.13) Newton proposed

$$q_x = h\,(T_S - T_A) \qquad (5.14)$$

where h is the heat transfer coefficient. We will find that this is much easier to correlate than the film thickness δ. Clearly, h will be a function of the flow as well as the properties of the fluid. This feature characterizes and distinguishes convective heat transfer.

Forced Convection

Forced convection refers to the situation where the fluid is forced past the surface by some external means such as a pump. The flow can be either internal, as inside a conduit, or external, as past an object.

It is customary to correlate forced convective heat transfer coefficients in terms of certain fundamental dimensionless groupings of the physical variables which characterize the flow situation and the fluid properties. These groups are summarized as follows:

The *Reynolds number* (*Re*) is a dimensionless group that characterizes the flow, either inside conduits or around objects:

$$Re = D\,v\,\rho/\mu \tag{5.15}$$

where

D = characteristic diameter or length, L
v = characteristic fluid velocity, L/t
ρ = mass density of the fluid, M/L^3
μ = viscosity of the fluid, M/Lt

In general, the higher the Reynolds number, the more turbulent or chaotic the flow situation. For example, for steady flow inside long cylindrical tubes, a Reynolds number of 2,100 represents the upper limit of stable smooth, streamlined laminar flow, while a Reynolds number of 4,200 represents the lower limit of turbulent flow and the appearance of eddies and local fluctuations in the velocity. (The Reynolds number in the aorta is about 2,500 and in the smaller capillaries is on the order of 0.10.) The Reynolds number can also be shown to be proportional to the relative magnitudes of the inertial forces associated with the fluid and the viscous forces of the fluid flow. For flow past spheres, a Reynolds number of 0.10 represents the upper limit of Stokes' law flow, that is, where the drag force can be predicted exactly by the well-known expression $6\pi v \mu R$.

The *Prandtl number* (*Pr*) is a dimensionless group characterizing the properties of the fluid. It is defined as

$$Pr = \hat{C}_p\,\mu/k \tag{5.16}$$

where

$\hat{C}_p$ = the fluid heat capacity, Q/MT
μ = fluid viscosity, M/LT
k = the fluid thermal conductivity, Q/LTt

On closer examination the Prandtl number will be seen to be the ratio of the kinematic viscosity to the thermal diffusivity—a measure of how well the fluid conducts momentum relative to how well it conducts heat.

It is convenient also to group the heat transfer coefficient h into a dimensionless group, the *Nusselt number* (Nu) defined as

$$Nu = h\,D/k \tag{5.17a}$$

One interpretation of the Nusselt number can be obtained by comparing (5.13) and (5.14). From these we see that $h = k/\delta$. Putting this into (5.17a),

$$Nu = D/\delta \tag{5.17b}$$

which indicates that the Nusselt number can be thought of as an equivalent film thickness. That is, high Nusselt numbers indicate a low film thickness in relation to the characteristic dimension.

The orders of magnitude of forced convective heat transfer coefficients are given in Table 5.2.

Table 5.2. Ranges of Values of Forced Convective h's

Gases	10–100 kcal/m^2 hr °C
Viscous fluids	50–500 kcal/m^2 hr °C
Water	1000–100,000 kcal/m^2 hr °C

Table 5.3 summarizes some typical correlations used to estimate values of h for forced convection problems. For additional correlations, the works listed at the end of this chapter may be consulted.

Table 5.3. Typical Correlations for h's

Inside tubes:	
$Re > 20{,}000$	$Nu = 0.023\, Re^{0.8}\, Pr^{0.4}$
Laminar flow	$Nu = 1.86\, (RePrD/L)^{0.33}$
Around objects:	
Spheres	$Nu = 2.0 + 0.60\, Re^{1/2}\, Pr^{1/3}$
Cylinders	$Nu = (0.35 + 0.56\, Re^{1/2})\, Pr^{0.3}$

Some examples of the use of these coefficients are appropriate at this point.

Example 5.4. Cooling of a Thin Plate

In this example we consider the rate of cooling of a very thin metal plate suspended as shown in Figure 5.4.

If the plate is initially at $T_o = 100$°C throughout, how long will it require to cool to the ambient temperature? What will be its temperature after one minute if the applicable value of the convective heat transfer coefficient at the surface is 25 kcal/m^2 hr °C, the mass is 1 kgm, the heat capacity is 0.50 cal/gm °C, and the area A is 0.5 m^2?

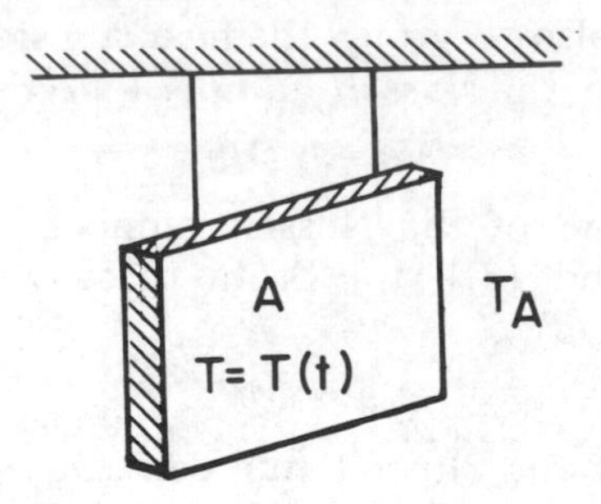

Fig. 5.4. Transient cooling of a heated plate.

If the plate is very thin and has a high thermal conductivity, then the following dimensionless number, the *Biot number*, is much less than unity.

$$Biot = h\,\Delta x/k << 1 \tag{5.17c}$$

In this case, the temperature gradient within the plate will be much less than the gradient in the film, and the internal differences can be ignored. Then we can apply a closed-system energy balance directly to the entire plate. The change in internal energy will be equal to the heat loss.

$$dU/dt = \dot{Q}$$

$$m_{tot}\,\hat{C}_p\,dT/dt = -h2A(T - T_A) \tag{5.18}$$

Rearranging and integrating and using the initial condition,

$$\int_{T_o}^{T} \frac{dT}{T - T_A} = -\int_0^t \frac{h2Adt}{m_{tot}\,\hat{C}_p} \tag{5.19}$$

$$\frac{T - T_A}{T_o - T_A} = \exp(-2hAt/m_{tot}\,\hat{C}_p) \tag{5.20}$$

$$T = 25 + 75\exp(-0.83\,t)$$

and after one minute the temperature is 58°C. The heat transfer coefficient h plays an important role in the time constant of this process.

Example 5.5. Cooling of a Fluid Flowing in a Tube

Figure 5.5 depicts a common engineering problem. Fluid at temperature T_o enters a tube with the walls held at a constant temperature T_W by some external means. The relationship between the fluid temperature T and the length L is desired for a fluid of known physical properties and a fixed flow rate. We can assume for simplicity that the fluid flows through the tube as a plug; that is, its properties

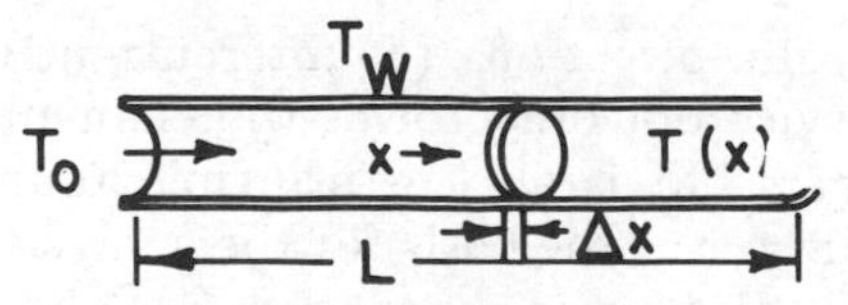

Fig. 5.5. Cooling of a fluid flowing in a tube.

exhibit insignificant radial variation. The convective heat transfer coefficient at the wall is denoted by h.

An appropriate way to solve this problem is to perform an energy balance on a disc of space at a distance x down the pipe. The disc will have a thickness Δx. Using (4.4),

$$w\,\hat{H}_{\text{in}} - w\,\hat{H}_{\text{out}} = -\dot{Q} \tag{5.21}$$

$$w\,\hat{C}_p\,T|_x - w\,\hat{C}_p\,T|_{x+\Delta x} = h\pi D\Delta x\,(T - T_w) \tag{5.22}$$

We have assumed that we can represent the temperature and velocity of the fluid entering and leaving our disc by the average values T and w. Dividing through by Δx and taking the limit as Δx goes to zero,

$$-\,w\,\hat{C}_p\,dT/dx = \pi Dh\,(T - T_W) \tag{5.23}$$

Since $w = \rho v\pi D^2/4$,

$$\int_{T_o}^{T} \frac{dT}{T - T_W} = -\frac{4h}{\hat{C}_p\,\rho\,vD}\int_0^L dx \tag{5.24}$$

$$\ln\frac{T - T_W}{T_o - T_W} = -\frac{4\,h\,L}{\rho\,C_p\,v\,D} \tag{5.25}$$

or, on rearranging,

$$\ln\frac{T - T_W}{T_o - T_W} = -\frac{Nu}{Re\,Pr}\,\frac{4\,L}{D} \tag{5.26}$$

It is seen that the right side is a grouping of previously mentioned dimensionless groups and is in itself a well-known dimensionless group in convective heat transfer, the *Graetz number.* The accomplished temperature change is seen to be a logarithmic, or exponential, function of length.

Example 5.6. Double Pipe Heat Exchanger

Figure 5.6 depicts a common piece of heat transfer equipment, the double pipe heat exchanger. The object is to exchange heat from the fluid stream flowing inside the inner pipe to the other fluid stream flowing in the outer annular space formed by the larger outside pipe. The fluids may be flowing in the same direction (concurrently) or in opposite directions (countercurrently). This type of heat-exchanging device not only forms the elements of many industrial heat exchangers but is also in principle found in physiology. Equation (4.4) again forms the basis for the analysis. For each fluid,

$$w\,\hat{C}_p\,T_{\text{in}} - w\,\hat{C}_p\,T_{\text{out}} = -\dot{Q} \tag{5.27}$$

Since one fluid picks up all the heat given up by the other (in the absence of losses),

$$w_i\,\hat{C}_{pi}\,(T_{i1} - T_{i2}) = w_o\,\hat{C}_{po}\,(T_{o2} - T_{o1}) \tag{5.28}$$

For example, if

$$w_i = 10 \text{ kgm/sec},\ w_o = 5 \text{ kgm/sec},\ C_{pi} = C_{po} = 1 \text{ kcal/kgm }^\circ\text{C}$$

$$T_{o2} - T_{o1} = 2(T_{i1} - T_{i2}) \tag{5.29}$$

Note that all but one temperature has to be specified, since only one independent relation is available at this point. Also, (5.28) tells us nothing about the size of the exchanger needed to accomplish the heat transfer required in (5.27). This information is contained in the rate expression, that is, in the application of Fourier's law and the appropriate heat transfer coefficients for the situation. Careful consideration of Figure 5.6 will reveal that there are three resistances to

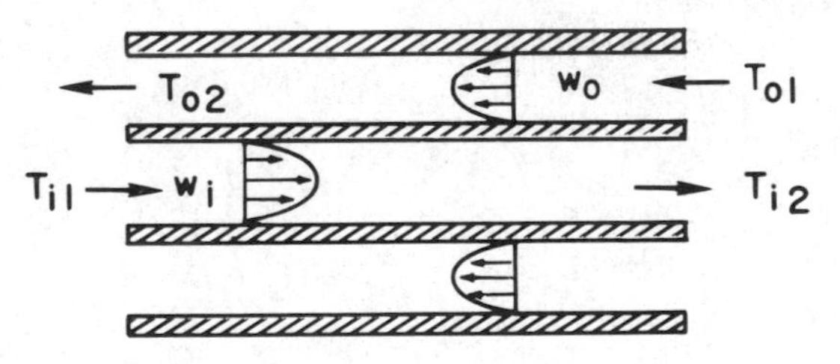

Fig. 5.6. Double pipe heat exchanger.

be overcome in transferring the heat from one fluid to the other in this apparatus. They may be described as follows:

1. The film of slowly moving fluid on the outside of the inside pipe.
2. The wall of the inner pipe.
3. The film of slowly moving fluid on the inside of the inner pipe.

We have seen in Example 5.1 that it is possible to consider resistances to heat transfer in an analogous fashion to electrical resistances and to sum them to obtain an overall resistance when they occur in series. Such a situation exists here. We employ heat transfer coefficients for the films inside and outside the inner pipe and define an overall coefficient U_o as

$$1/U_o = 1/h_o + \Delta x\, D_o/k_m\, D_m + D_o/h_i\, D_i \tag{5.30}$$

where

U_o = overall heat transfer coefficient, based on the outer area of the inner pipe

h_o = convective heat transfer coefficient for the outside of the inner pipe

Δx = thickness of the pipe wall

k_m = thermal conductivity of the pipe wall

D_o = outside diameter of the inner pipe

D_i = inside diameter of the inner pipe

h_i = convective heat transfer coefficient for the inside of the inner pipe

D_m = mean diameter of the inner pipe

The overall heat transfer coefficient could easily be based on the inside area of the inner pipe, in which case (5.30) would be

$$1/U_i = D_i/D_o\ h_o + \Delta x\ D_i/k_m\ D_m + 1/h_i \tag{5.31}$$

The overall coefficient can now be used in the overall rate equation to compute the total amount of heat exchanged between the two fluids.

$$\dot{Q} = U_o\ A_o\ \Delta T_{LM} = U_i\ A_i\ \Delta T_{LM} \tag{5.32}$$

where

$$A_o = \pi\ D_o\ L$$
$$A_i = \pi\ D_i\ L$$

The logarithmic mean temperature difference ΔT_{LM} at the two ends of the exchanger is defined for this countercurrent exchanger as

$$\Delta T_{LM} = \frac{(T_{i1} - T_{o2}) - (T_{i2} - T_{o1})}{\ln \dfrac{(T_{i1} - T_{o2})}{(T_{i2} - T_{o1})}} \tag{5.33}$$

For concurrent flow the definition requires only that the 1 and 2 subscripts on either of the two phases be switched. It is necessary to use a logarithmic average driving force rather than an arithmetic mean because, as seen in Example 5.5, the variation of temperature with length is exponential (or logarithmic).

Now we have two equations to relate the temperatures in the exchanger and the size of the exchanger, (5.28) and (5.32). In summary

$$\dot{Q} = w_i\ \hat{C}_{pi}(T_{i1} - T_{i2}) = w_o\ \hat{C}_{po}\ (T_{o2} - T_{o1})$$
$$= U_o\ A_o\ \Delta T_{LM} = U_i\ A_i\ \Delta T_{LM}$$
$$1/U_o = 1/h_o + \Delta x\ D_o/k_m\ D_m + D_o/h_i\ D_o$$

$$A_o = \pi D_o L$$
$$h_o = h_o (Re_o, Pr_o)$$
$$h_i = h_i (Re_i, Pr_i) \tag{5.34}$$

These equations represent the design equations for the double pipe heat exchanger. The individual coefficients must be obtained from available correlations, some sources of which are listed at the end of the chapter.

Both the mechanisms of convection (in the fluid) and conduction (in the pipe wall) play an important role in the heat transfer. This is usually the case in heat transfer equipment problems, and the designer must often evaluate the relative importance of the various mechanisms in his treatment of the problem.

Free Convection

Free convection, or natural convection, differs from forced convection in that the stimulus for the fluid motion, rather than an externally imposed pressure, comes from buoyancy forces in the fluid usually resulting from a gradient in density. The density gradient in turn usually results from the presence of a temperature gradient in a compressible fluid. Hot gases rising from a smokestack are a good example of such an effect.

The use of heat transfer correlations for free convection problems to predict the film coefficients is quite analogous to the situation in forced convection. The major difference comes in the fundamental grouping of variables which replaces the Reynolds number and characterizes the buoyancy forces present in the situation. This quantity is the *Grashof number*, denoted Gr, and defined

$$Gr = D^3 \rho^2 g \beta \Delta T / \mu^2 \tag{5.35}$$

where

D = characteristic length of the object
ρ = mass density of the fluid
g = local acceleration of gravity
β = coefficient of volume expansion ($1/T$ for an ideal gas)
ΔT = characteristic temperature difference
μ = fluid viscosity

Table 5.4 gives some order of magnitude values, and Table 5.5 some typical correlations for free convective heat transfer coefficients. If the Grashof number is subjected to close scrutiny, the reader should be able to see that it is quite similar to a Reynolds number squared.

Table 5.4. Range of Magnitudes for Free Convective Heat Transfer Coefficients

Gases	3 - 20 kcal/m^2 hr °C
Liquids	100 - 600 kcal/m^2 hr °C

Table 5.5. Correlations for Free Convective Heat Transfer Coefficients

Sphere	$Nu = 2 + 0.6\,Gr^{1/4}\,Pr^{1/3}$
Horizontal cylinder	$Nu = 0.525(GrPr)^{1/4}$
Vertical plate	$Nu = 0.590(GrPr)^{1/4}$
Vertical cylinder	$Nu = 0.13(GrPr)^{1/3}$

Example 5.7. Convective Cooling of a Sphere

This example will attempt to elucidate the relation between the three modes of heat transfer discussed so far—conduction, forced convection, and free convection. Example 5.2 has shown how the internal temperature of a cooling sphere varies with time if the surface is held at constant temperature. A more realistic problem is the consideration of the cooling of a heated sphere when the surface cannot be held at a constant temperature and the rate of cooling is limited by the resistance of the surface film, that is, there is a resulting temperature profile in the air around the sphere, as shown in Figure 5.7. Usually the temperature difference $T_R - T_a$ is much

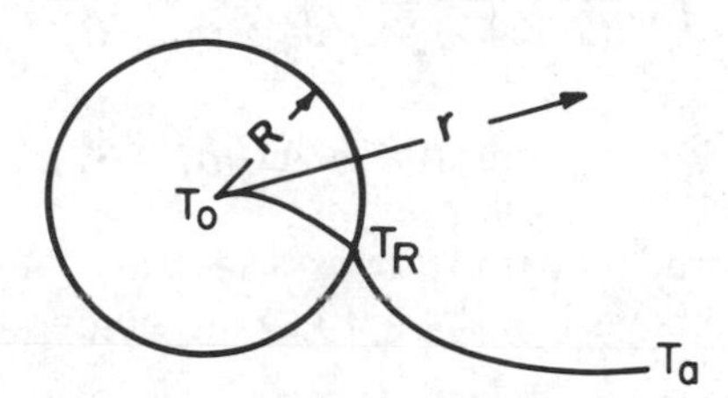

Fig. 5.7. Convective cooling of a heated sphere.

greater than the difference $T_o - T_R$, so that the significant resistance to heat transfer is in the film around the sphere.

If the fluid around the sphere is completely stagnant, then the heat will be transferred exclusively by conduction, and (5.5c) can be used to determine the limiting value of the Nusselt number at this condition. If $T = T_R$ at $r = R$, and $T = T_a$ at a very large value of r, then (5.5c) becomes

$$q_o = -\frac{k}{R^2}\frac{(T_R - T_a)}{-1/R} = k(T_R - T_a)/R \tag{5.36}$$

Since $q_o = h(T_R - T_a)$ by definition,

$$h = k/R \tag{5.37}$$

so that the Nusselt number becomes

$$Nu = hD/k = 2 \tag{5.38}$$

If forced convection around the sphere begins, then the film coefficient and the ability to transfer heat increases, as predicted from Table 5.3,

$$Nu = 2 + 0.6\, Re^{1/2}\, Pr^{1/3} \tag{5.39}$$

while if natural convection occurs, the increase is predicted from Table 5.5 as

$$Nu = 2 + 0.6\, Gr^{1/4}\, Pr^{1/3} \tag{5.40}$$

In both cases all the fluid properties are evaluated at an average film temperature T_f, where

$$T_f = (T_R + T_a)/2 \tag{5.41}$$

If the rate of cooling of the sphere is desired, application of the closed-system energy balance expression gives

$$\dot{T} = \dot{Q} \tag{5.42}$$

$$m_{\text{tot}}\, \hat{C}_p\, dT/dt = -4\pi R^2\, h\, (T - T_a) \tag{5.43}$$

$$\int_{T_o}^{T} \frac{dT}{(T - T_a)} = - \int_0^t \frac{h\, A\, dt}{m_{\text{tot}}\, \hat{C}_p} \tag{5.44}$$

$$\frac{T - T_a}{T_o - T_a} = \exp\,(-h\, At/m_{\text{tot}}\, \hat{C}_p) \tag{5.45}$$

where h is considered constant over the time of the cooling, and temperature variations within the sphere are neglected.

RADIATIVE HEAT TRANSFER

Energy transport by radiation is distinctly different from the mechanisms discussed earlier in this chapter. Electromagnetic radiation originating from excited molecules returning to lower energy states in material at a high temperature causes a net amount of thermal energy to be transferred to surrounding surfaces of lower temperatures. From a corpuscular point of view, radiant energy can be thought of as energy carried by photons, particles having only energy content bereft of mass and charge. The energy content of a photon is given by the product $h\nu$, where h is Planck's constant

$$h = 6.62 \times 10^{-27} \text{ erg-sec}$$

and ν is the frequency of the emitted radiation. The frequency and wavelength are related to the velocity of propagation by

$$\nu\lambda = c \tag{5.46}$$

where

λ = wave length, L
ν = frequency, $1/t$
c = speed of light (3×10^{10} cm/sec)

From a macroscopic viewpoint if we consider the amount of energy impinging on a solid surface, we can define the *absorptivity* a of the surface as the fraction of the incident energy which is absorbed. We can also define a_λ, the *monochromatic absorptivity*, as the fraction of incident energy of a given frequency that is absorbed. For any real substance a and a_λ are less than one, and a_λ is a function of frequency.

It is convenient to define a body that is a perfect absorber, that is, one for which $a = a_\lambda = 1.0$ for all frequencies, and refer to such a body as a *black body*. For engineering purposes, it is also useful to define another theoretical body for which a_λ is less than one but is constant for all frequencies. Such a body is called a *gray body*.

In a similar way, the *emissivity* e of a substance is defined as the fraction of the energy that would be emitted by a black body at a given temperature that is actually emitted by the substance. The *monochromatic emissivity* e_λ can be similarly defined. Kirchkoff's law states that at radiative equilibrium the absorptivity and the emissivity at any frequency must be equal.

While not important for our purposes, it is probably instructive to define the *reflectivity* r as the fraction of the incident radiation which is reflected, and the *transmissivity* t as the fraction which is transmitted ($t = 0$ for opaque materials). Then the sum of the absorptivity, reflectivity, and transmissivity must equal one, that is, $a + r + t = 1$ for any substance.

Figure 5.8 shows how the energy emitted from heated bodies varies with wavelength for black, gray, and real bodies and how the emitted energy for a black body varies with temperature. The black body curves in Figure 5.8 may be described by the Planck distribution law.

$$q^e_{b\lambda} = \frac{2\pi\, c^2\, h}{\lambda^5}\left[\frac{1}{\exp(ch/\lambda\, kT) - 1}\right] \tag{5.47}$$

where k = Boltzmann's constant

If the areas under the black-body curves are evaluated either

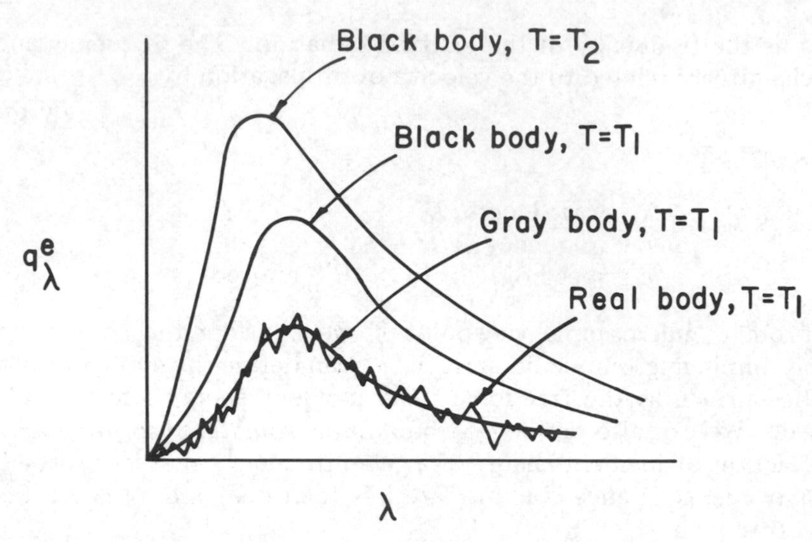

Fig. 5.8. Radiant energy spectra.

experimentally or analytically, it is found that the law of Stefan-Boltzmann results.

$$q_b^e = \sigma T^4 \tag{5.48}$$

where

σ = Stefan-Boltzmann constant = 0.1712×10^{-8} BTU/hr ft^2 °R^4
= 81.3×10^{-8} cal/min m^2 °K^4

T = absolute temperature

For nonblack bodies,

$$q^e = e\,\sigma\,T^4 \tag{5.49}$$

is a good approximation, where *e* is the average emissivity. Table 5.6 gives some typical values of *e* for real materials.

Table 5.6. Typical Emissivity Values

Substance	*Temperature*	*Emissivity*
Water	0°C	0.95
Water	100°C	0.96
Ice	0°C	0.86
Skin	37°C	0.87
Glass	25°C	0.93
Plaster	25°C	0.91
Silver	25°C	0.02
Black lacquer	25°C	0.80
White lacquer	25°C	0.80

Further observation of Figure 5.8 shows that the wavelength at which the maximum energy is emitted shifts with temperature. This shift is described by Wien's displacement law:

$$\lambda_{max} T = 0.2884 \text{ cm } ^\circ\text{K} \tag{5.50}$$

This law, which can be derived directly by differentiating (5.47), is the basis for the operation of the optical pyrometer, a device used to measure the temperature of remote objects by measuring the wavelength at which the radiated energy is a maximum.

The net radiant heat transferred between two black surfaces with temperatures T_1 and T_2 can be written using the Stefan-Boltzmann law as

$$\dot{Q}_{12} = A_1 \sigma F_{12} (T_1^4 - T_2^4) \tag{5.51}$$

where A_1 is the area of surface 1, and F_{12} is the view factor. This quantity represents the fraction of the total energy emitted by body 1 that is intercepted by body 2. It is exclusively a function of the geometry and orientation of the two bodies. For example, for two infinite, plane parallel walls $F_{12} = 1$. For two discs of radius R separated by a distance H and parallelly facing each other, it is

$$F_{12} = \frac{1 + 2B^2 - \sqrt{1 + 4B^2}}{2B^2} \qquad B = R/H \tag{5.52}$$

If the two discs are very far apart, $F_{12} = R^2/H^2$; whereas if they are very close, $F_{12} = 1 - R/H$.

Example 5.8. Energy Flux from the Sun

Given the following data, our task is to compute the energy flux which the earth is receiving from the sun.

Sun data:

$\sigma T_1^4 = 2 \times 10^7$ BTU/hr ft^2 (from optical pyrometer measurements)

$D_1 = 8.6 \times 10^5$ miles (diameter of the sun)

$H = 9.3 \times 10^7$ miles (distance from sun to earth)

The radius of the earth will be taken as 8.1×10^3 miles.

If we consider the sun and the earth as two large discs very far apart, for view purposes, then $F_{12} = R_2^2/H^2$, and the flux hitting the earth will be

$$\dot{Q}_{12} = \frac{\sigma A_1 F_{12} T_1^4}{\pi R_2^2} = \sigma \times \frac{D_1^2}{4} \times \frac{T_1^4}{H^2} = 430 \text{ BTU/hr ft}^2 \tag{5.53}$$

The view factor often has to be approximated.

If the radiating bodies are nonblack, often they may be modeled as gray bodies. (In the last example both the sun and the earth may be considered as black without significant error.) Then a reasonable treatment is to use

$$\dot{Q}_{12} = \sigma A_1 F_{12} (e_1 T_1^4 - a_1 T_2^4) \tag{5.54}$$

where a_1 is estimated as the value of e_1 at temperature T_2.

If there are several surfaces involved but only two are transferring heat—that is, the others are all adiabatic or refractory surfaces which are interconnecting surfaces 1 and 2—then the corrected view factor is

$$\overline{F}_{12} = F_{12} + F_{1R} \left[\frac{F_{R2}}{(F_{R1} + F_{R2})} \right] \tag{5.55}$$

where the subscript R indicates the refractory surfaces. As an example, two parallel squares of sides S separated by a distance S have a view factor $F_{12} = 0.40$. When they are interconnected by refractory surfaces, the view factor F_{12} is 0.68; that is, the heat transfer between them increases by 70% due to the interconnecting adiabatic surfaces.

If the two interchanging surfaces are gray, then the view factor must be further corrected by

$$1/A_1 \mathfrak{F}_{12} = 1/A_1 \overline{F}_{12} + (1/e_1 - 1)/A_1 + (1/e_2 - 1)/A_2 \tag{5.56}$$

and used in the transfer equation as

$$\dot{Q}_{12} = A_1 \mathfrak{F}_{12} \sigma (T_1^4 - T_2^4) \tag{5.57}$$

Another useful relation for view factors is

$$A_1 F_{12} = A_2 F_{21} \tag{5.58}$$

Radiative heat transfer problems that involve real materials for all but the simplest geometries become increasingly complex. For this reason the use of radiative heat transfer coefficients often becomes necessary or desirable. Such a coefficient can be defined by the relation

$$\dot{Q}_{12} = h_r A_2 (T_1 - T_2) \tag{5.59}$$

It is tacitly assumed here that the fourth power temperature difference may be approximated, at least for small temperature differences, by the simple difference. If (5.57) and (5.59) are compared, it is seen that

$$h_r = A_1 \quad {}_{12}\, \sigma (T_1^2 + T_2^2)(T_1 + T_2)/A_2 \tag{5.60}$$

The radiative heat transfer coefficient is seen to be a function of the area ratio, the view, the emissivities of the surfaces, and the temperatures of the two bodies. For situations such as heat transfer from people, where the temperatures are not far apart on an absolute basis, the use of such coefficients is particularly appropriate.

Since radiative heat transfer takes place between surfaces, the temperature of the intermediate or ambient air does not materially affect this mode of transfer. This makes it possible, for instance, for water to freeze when exposed to the night sky despite ambient air temperatures well above the freezing point. The radiative transfer to the sky, which has a very low effective black body temperature, can often overcome the convective heating from the ambient air. Another example of the effect of surface radiation occurs in air-conditioned buildings. Most of the cooling experienced by people in these buildings comes from radiative loss to the subcooled surfaces rather than to the ambient air. Chapter 6 will discuss this in more detail.

Table 5.7. Key Heat Transfer Relations

Method	*Equations*
Conduction:	
A. Steady-state–no generation	5.4a–c, 5.5a–c
B. Steady-state–internal generation	5.11a–c
C. Transient–no generation	5.10
Convection:	
A. Forced convection	5.14, 5.15, 5.16, 5.17
B. Transfer coefficient correlations	Tables 5.3, 5.5
C. Grashof number	5.35
D. Heat exchanger design	5.34
Radiation:	
A. Interchange between black bodies	5.51
B. Interchange between gray bodies	5.54
C. Interchange between black bodies with connecting refractory surfaces	5.55
D. Interchange between gray bodies with partial view or refractory surfaces	5.56, 5.57
E. Radiative heat transfer coefficient	5.59, 5.60

SUMMARY

In this chapter we have presented the basic principles of heat transfer, and have discussed in some detail the three major mechanisms of thermal energy transfer.

Bibliography

Many excellent references are available for further study and as a source of correlations and experimental results for the general topic of heat transfer. Four good examples are:

Bird, R. B., Stewart, W. E., and Lightfoot, E. N. *Transport Phenomena*. Wiley, 1960.

Carslaw, H. S., and Jaeger, J. C. *Conduction of Heat in Solids*, 2nd ed. Oxford, 1959.

McAdams, W. H. *Heat Transmission*, McGraw-Hill, 1954.

Perry, J. H. *Chemical Engineer's Handbook*. McGraw-Hill, 1968.

chapter 6

Heat Transfer in Living Systems

INTRODUCTION

In this chapter we shall apply the principles developed in the preceding chapter to the study of heat transfer in and from living systems. We have previously set the framework for this by developing the general overall energy balances for both closed and open systems in Chapters 3 and 4. We have had to postpone the detailed consideration of energy problems until we had developed the facility for making computations of actual rates of heat transfer away from the systems in question. All the modes of heat transfer discussed in Chapter 5 play major roles in sustaining life in biological systems. In addition, problems arising in the design and operation of life support systems and artificial organs require the facility for predicting the relations between heat transfer area, geometry, temperature, fluid properties, environmental parameters, and the other operating parameters of the system.

In this chapter we will first devote our attention to an analysis of the rates of heat transfer from the living system to the environment. Then we shall examine several aspects of heat transfer within the system and discuss the role of thermal energy transfer in the thermoregulatory systems of the body.

HEAT TRANSFER TO THE ENVIRONMENT

Our first consideration will be the detailed discussion of the heat loss term $\dot{Q}'$, which has occurred in the overall energy balance for living subjects, such as (4.16)

$$\dot{M}_o + \Delta \dot{M} + \dot{T} = \hat{H}_f \, w_f - \dot{Q}_{\text{resp}} - \dot{Q}_{\text{vap}} - \dot{Q}' - \dot{W} \tag{6.1}$$

All three modes of heat transfer are important for this application, so we will divide $\dot{Q}'$ into three parts and discuss each part separately, as

$$\dot{Q}' = \dot{Q}_{\text{conv}} + \dot{Q}_{\text{cond}} + \dot{Q}_{\text{rad}} \tag{6.2}$$

It is useful to summarize the normal relative magnitudes of the various mechanisms for heat transfer, so that some idea of the relative importance of these terms might be gained. For an average, resting, clothed adult person exposed to an ambient temperature of 70°F, the approximate relative percentages of the energy loss terms are:

$$\dot{Q}_{\text{rad}} \sim 60\% \text{ of } \dot{M} \qquad \dot{Q}_{\text{resp}} \sim 10\%$$

$$\dot{Q}_{\text{conv}} \sim 20\% \qquad \dot{Q}_{\text{vap}} \sim 10\%$$

$$\dot{Q}_{\text{cond}} \sim 0$$

As indicated, the normal conductive heat transfer loss away from the body is very small, but conduction plays a major role inside the body, along with forced convection. Since in almost all instances this mode is negligible for external losses, our discussion in this section will be quite limited. Perhaps the most dramatic example of conductive loss occurs when one stands barefoot on a cold bare floor. The reduction of thermal conductivity experienced when stepping from the bare floor onto a carpeted floor is a dramatic illustration of Fourier's law.

The subject of convective and radiative heat exchange from the human body continues to be of great research interest. In this section we will present some of the major results and concentrate on the physical principles involved.

The use of heat transfer coefficients considerably simplifies the calculation of heat transfer rates from people. The coefficients for convective transfer depend greatly on posture and the type of clothing being worn, but in general they follow the forms indicated in Table 6.1.

Table 6.1. Typical Values of Convection Coefficients

Situation	h_c (kcal/m^2 hr °C)
Free convection, nude, standing	2.3
Free convection, nude, seated	2.0
Forced convection, standing	$7.5v^{0.67}$ (v = m/sec)
Forced convection, seated	$6.4v^{0.67}$

The heat flow by convection can then be computed by

$$\dot{Q}_{\text{conv}} = h_c A_c (T_s - T_a) \tag{6.3}$$

where

A_c = area of the convecting surface
T_s = average skin (or surface) temperature
T_a = temperature of the ambient air

The average skin temperature may be computed in a number of ways. One popular method is the use of a relation such as

$$T_{\text{skin}} = 0.07\ T_{\text{head}} + 0.14\ T_{\text{arms}} + 0.05\ T_{\text{hands}} + 0.07\ T_{\text{feet}} + 0.13\ T_{\text{legs}} + 0.19\ T_{\text{thighs}} + 0.35\ T_{\text{trunk}} \tag{6.4}$$

Under neutral conditions, where the body is not subject to either positive or negative thermal stress, the average skin temperature is about 34.2°C.

The area for convection A_c must be estimated as some fraction of the total body surface area (about 1.8 m^2) for the situation being encountered.

Example 6.1. Equivalent Free-Convection Ambient Temperature

Suppose that the entire metabolic energy load had to be dissipated exclusively by free convection. What ambient temperature would be required for a nude standard man?

The standard $\dot{M}_o$ is - 72 kcal/hr. If this were all to be lost by free convection,

$$-\dot{M}_o = \dot{Q}' = \dot{Q}_{\text{conv}} = h_c\ A_c\ (T_s - T_a)$$

$$T_s - T_a = 72\ \text{kcal/hr}/2.3\ \text{kcal/m}^2\ \text{hr}\ °\text{C}\ (1.8\ \text{m}^2)$$

$$T_s - T_a = 17.4°\text{C}$$

If

$$T_s = 34.2$$

then

$$T_a = 16.8°\text{C}\ (63°\text{F}) \tag{6.5}$$

At temperatures below this, the body would begin to cool. Remember that this neglects all radiative or evaporative losses and is for a completely nude individual doing no work.

Table 6.1 also indicates how drastically the total convective loss increases with ambient air velocity. An ambient air velocity of only 1 m/sec (2.2 mi/hr) increases the loss rate over that of free convection by the ratio of 7.5/2.3, or 3.25, assuming that the areas involved for heat transfer in the two situations are roughly equivalent. This would work out to a neutral temperature in Example 6.1 of 32.9°C (91°F) if all the generated heat were removed from a nude man solely by these two mechanisms. An ambient air velocity of only 2 m/sec increases the loss rate by a factor of 5.2. We will see that the effect of air velocity on clothed persons is not so drastic.

The subject of heat losses from human subjects immersed in water has received much attention. Based on work done by several researchers, a representative value for the overall heat transfer coefficient for this situation is 16.5 kcal/m^2 hr °C, based on the temperature difference between the body core and the water.

Example 6.2. Equivalent Water Immersion Temperature

Using this information, we can compute an equivalent temperature for water immersion that will exactly balance the resting metabolic energy rate, as we did in Example 6.1 for natural convection in air. Following a similar procedure, for a standard man resting and totally immersed in water,

$$\dot{Q}_{H_2O} = h_w\ A_w\ (T_c - T_w)$$

$$T_c - T_w = 72\ \text{kcal/hr}/16.5\ \text{kcal/m}^2\ \text{hr}\ ^\circ\text{C}\ (1.8\ \text{m}^2)$$

$$T_c - T_w = 2.4^\circ\text{C} \quad T_w = 34.6^\circ\text{C}\ (94^\circ\text{F}) \tag{6.6}$$

At water temperatures below this the body would begin to cool, while at temperatures above it the body would begin to warm.

Comparison with the corresponding result in Example 6.1 demonstrates the superior heat transfer properties of water as compared to air. These are mainly the high thermal conductivity and high heat capacity. The temperature drop required in water is only one-seventh that for air.

Example 6.3. Exercise and Water Immersion

Suppose the subject described in Example 6.2 exercises vigorously while immersed. What is the relationship between the neutral water temperature and exercise rate if all the energy is dissipated as heat?

We resort to a closed-system energy balance,

$$\dot{T} + \dot{M} = -\dot{Q}' - \dot{W} \tag{6.7}$$

which shows that the required heat loss for $\dot{T} = 0$ is directly related to metabolic activity. Therefore, the required temperature difference will also be directly proportional to metabolic activity if we assume that the convective heat transfer coefficient is unaffected by exercising and that the energy lost by work output is small compared to the other two terms (probably not a very good assumption for some types of exercise). Then, for water,

$$(T_c - T_w)/2.4 = \dot{M}/\dot{M}_o \tag{6.8a}$$

and for air,

$$(T_c - T_a)/17.4 = \dot{M}/\dot{M}_o \tag{6.8b}$$

For a metabolic energy conversion rate of five times normal, $T_w = 25°C$ (77°F) and $T_a = -52°C$ (-62°F). Although these equivalent temperatures are informative, it should be stressed that particularly in the air case their actual physical significance is very limited. It would be extremely difficult to construct a situation in air where only convective heat losses were significant. Also, these computations do not take forced convection into account. These values, however, are illustrative of the relation between free convection and metabolic rate.

The prediction of radiative losses from humans follows directly from the fundamentals presented in Chapter 5 with some empirical modifications. One expression commonly used is a slight modification of (5.54) and is

$$\dot{Q}_{rad} = \sigma e_w e_s (T_s^4 - T_w^4) A_r \tag{6.9}$$

where

e_w = emissivity of the surrounding walls
e_s = emissivity of the skin
T_w = surrounding wall temperature
T_s = average skin temperature
A_r = effective surface area for radiation. This quantity is often estimated as some fraction of the total body surface area. One popular value of the fraction is 0.77.

An alternative treatment is to use

$$\dot{Q}_{rad} = h_r A_r (T_s - T_w) \tag{6.10}$$

A representative value of h_r for nude humans would range from 3.5 to 5 kcal/m² hr °C.

Example 6.4. Relation of Neutral Ambient and Surface Temperatures

In this example we will prepare a plot which will show the relationship between the temperature of the surrounding surfaces and the ambient air temperature for a neutral situation, that is, where

$$-\dot{M}_o = \dot{Q}' = h_r A_r (T_s - T_w) + h_c A_c (T_s - T_a) \tag{6.11}$$

using the values

$$h_r = 5 \text{ kcal/m}^2 \text{ hr °C}$$
$$h_c = 2.3 \text{ kcal/m}^2 \text{ hr °C}$$
$$A_r = 1.4 \text{ m}^2$$
$$A_c = 1.7 \text{ m}^2$$
$$T_s = 35°C$$
$$\dot{M}_o = -72 \text{ kcal/hr}$$

Equation (6.11) can now be solved for T_a as a function of T_w to produce pairs of temperatures that will in combination provide a neutral environment. The results are

T_w (°C)	20.0	25.0	30.0	35.0	40.0
T_a (°C)	43.9	35.0	26.1	17.6	9.2

These arc plotted in Figure 6.1.

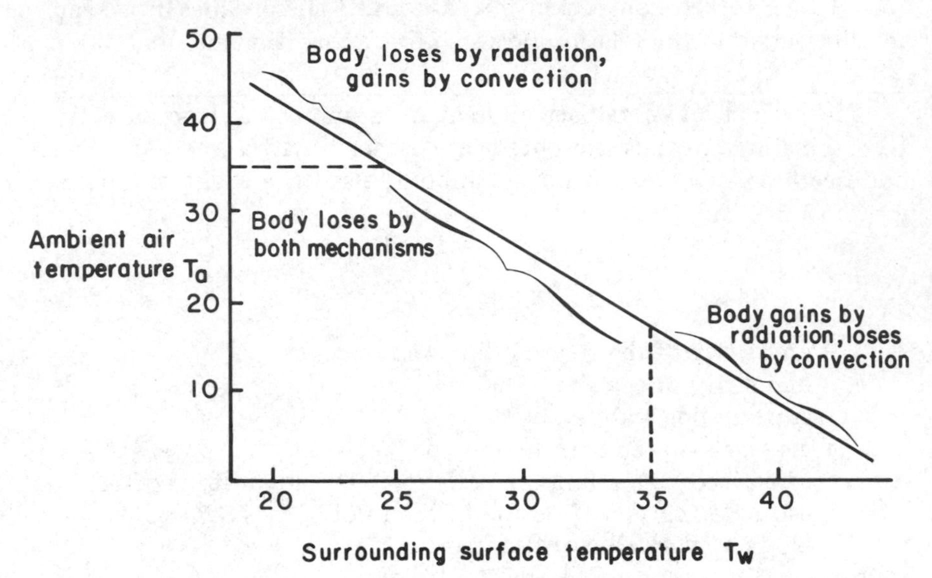

Fig. 6.1. Neutral temperature combinations.

This figure, which requires close examination, shows the relationship between the surface and air temperatures which will provide a neutral environment for a resting nude standard man. Similar balance lines could be plotted for various metabolic rates and values of the heat transfer coefficients. The reader should prepare his own such plots for a few cases of interest.

In the examples so far, and particularly in Example 6.4, we have neglected any changes in the average skin temperature. We shall see later that the thermal regulatory mechanisms of the body act to raise or lower the average skin temperature in an effort to increase or reduce heat loss in warm and cold environments. At that time we will describe the treatment of such changes and their effect on the total energy balance, but here we will continue to concentrate on the effects of environmental change.

For clothed individuals this type of treatment can be extended by defining an overall coefficient of heat transfer which sums the added resistances of the clothing layers—in a way similar to the treatment of the double pipe heat exchanger of Example 5.6 or as in the plane composite wall conduction problem of Example 5.1. We formulate the overall coefficient as follows, neglecting any changes in total area,

$$1/h_{tot} = 1/h_{sc} + 1/(h_{cc} + h_{cr}) \tag{6.12}$$

where

h_{tot} = overall heat transfer coefficient, skin to surroundings
h_{sc} = heat transfer coefficient, skin to clothing exterior
h_{cc} = convective heat transfer coefficient, clothing to surroundings
h_{cr} = radiative heat transfer coefficient, clothing to surroundings

Using a value for h_{sc} of 6.0 kcal/m^2 hr °C and clothing coefficients of h_{cc} and h_{cr} for still air, an overall coefficient in the range of 3–5 kcal/m^2 hr °C is representative. However, two points should be recalled here. First, the effects of changes in wind velocity must be accounted for, and second, the difference between surface and ambient temperatures must be allowed for. A resulting engineering approximation for the total rate of heat loss due to convection and radiation for a clothed person is

$$\dot{Q}'_{conv+rad} = 3.0(v/v_0)^{0.6}\, A_c(35 - T_a) + 2.0(35 - T_s)A_r \tag{6.13}$$

where

$$v_0 = 1 \text{ m/sec}$$

The variation of the overall convective coefficient with the 0.6 power of velocity is reminiscent of the correlations of Chapter 5 for forced convective heat transfer coefficients around objects. This value has been confirmed by many investigators.

Example 6.5. Prediction of Heat Transfer Coefficients Based on Correlations for Cylinders.

If we consider a human as a 1.5 ft diameter vertical cylinder, we can employ correlations of Chapter 5 for forced convection around cylinders to predict a value of h_c, and compare this with the experimental results from Table 6.1. From Table 5.3, we recall

$$Nu = (0.35 + 0.56Re^{1/2})\, Pr^{0.3} \tag{6.14a}$$

For air at 70° F,

$$Pr = 0.79$$

$$Pr^{0.3} = 0.93$$

$$\mu/\rho = 1.6 \times 10^{-4} \text{ ft}^2/\text{sec} = 1.5 \times 10^{-5} \text{ m}^2/\text{sec}$$

$$k = 0.015 \text{ BTU/ft °F hr} = 0.022 \text{ kcal/m °C hr}$$

given

$$D = 1.5 \text{ ft} = 0.46 \text{ m}$$

Substituting these values into the correlation (6.14a) and dropping the negligible nonvelocity dependent term, gives

$$h_c = 4.4\, v^{0.5} \text{ kcal/m}^2 \text{ hr } ^\circ\text{C} \tag{6.14b}$$

which compares with the result from Table 6.1;

$$h_c = 6.4\, v^{0.67} \tag{6.14c}$$

While the comparison is not exact, it does show the relationship and similarity between the two cases. For better agreement a larger value of D would be appropriate.

Example 6.6. Preparation of a Wind-Chill Chart

Equation (6.13) can be used as the basis for determining an equivalent ambient temperature to account for the effect of wind velocity. Such equivalent temperatures are the basis for wind-chill charts. These charts, popular in the winter in the Midwest, relate the equivalent temperature at zero wind velocity with the existing wind velocity and temperature. Neglecting radiative and evaporative losses, the solution of (6.13) for the two situations will give the necessary results. Table 6.2 gives a section of such a chart. The computation is carried out as

$$(v/v_0)^{0.6}\ (35 - T_a) = (35 - T_a^*) \tag{6.15}$$

where

v = ambient velocity in m/sec (0.448 m/sec/mph)
v_0 = 1 m/sec
T_a^* = equivalent temperature

Table 6.2. Equivalent Temperature T_a^*

T_a (°C)	*Wind Velocity, mph*		
	10	*20*	*30*
30	23	16	11
20	-2	-21	-36
10	-26	-58	-74
0	-51		

Table 6.2 shows that the effect of an appreciable wind velocity on a nude human, or for that matter on a lightly clothed human, is dramatic at low ambient temperatures.

It is sometimes convenient to treat the energy losses due to evaporation of water vapor from the skin as a heat loss term. It is possible to do this by defining a transfer coefficient for this situation and replacing the temperature driving force by a partial pressure driving force. Chapters 7 and 8 will treat this problem in detail, but for the present it will suffice to define $\dot{Q}_{vap}$ as

$$\dot{Q}_{vap} = h_v A_v (p_s - p_a) \tag{6.16}$$

where

h_v = heat transfer coefficient for evaporative transfer, kcal/m^2 hr mm Hg
A_v = area from which vaporization is taking place—the equivalent area of skin that is covered with water, m^2
p_s = partial pressure of water at the temperature of the skin included in A_v, mm Hg
p_a = partial pressure of water vapor in the ambient air, mm Hg

One must exercise considerable care in the application of (6.16). In particular, the evaluation of the wetted area A_v and the equivalent skin partial pressure require careful consideration. If the skin partial pressure is taken as the vapor pressure of water at body temperature (p_{H_2O} = 47 mm Hg), then the wetted area A_v will usually be only a very small fraction of the exposed skin area. Alternatively, if A_v is selected as the total exposed skin area, then it is necessary to use an adjusted value of p_s, that is, an estimated value much less than the vapor pressure. Cases do occur when the skin is completely covered with an evaporating water film, but in general these are rare.

A representative value for h_v based on measurements reported in the literature is $h_v = 11.9\, v^{0.6}$ kcal/m^2 hr mm Hg.

Example 6.7. Body Temperature Loss in a Newborn Infant

One situation where evaporative and convective losses are quite high is the birth and subsequent exposure of a nude, wet, newborn infant in the delivery room. In this example we will estimate the rate of body temperature loss for the period immediately following delivery, where the completely wet infant is temporarily exposed to the operating room environment. The following data are used:

Weight of infant: 5 lb_m (2.3 kg)
Surface area: 0.20 m^2
Basal metabolism: 2.0 kcal/kg hr
Room environment: 72° F, 55% relative humidity (p_{H_2O} = 11 mm Hg)

The basic relation for this case is

$$\dot{M} + \dot{T} = -\dot{Q}_{evap} - \dot{Q}_{rad} - \dot{Q}_{conv} \tag{6.17}$$

or, solving for $\dot{T}$,

$$\dot{T} = -\dot{M} - \dot{Q}_{evap} - \dot{Q}_{rad} - \dot{Q}_{conv} \tag{6.18}$$

and substituting,

$$m_{tot}\, \hat{C}_p\, dT/dt = -\hat{M}\, m_{tot} - h_v\, A_v(p_s - p_a) - h_r\, A_r(T_s - T_w) - h_c\, A_c(T_s - T_a) \qquad (6.19)$$

For the environment specified,

$$h_v = 11.9 \text{ kcal/m}^2 \text{ hr mm Hg}$$

$$h_c = 9.8 \text{ kcal/m}^2 \text{ hr } ^\circ\text{C}$$

$$h_r = 5.0 \text{ kcal/m}^2 \text{ hr } ^\circ\text{C}$$

and making the following area approximations,

$$A_v = 0.18 \text{ m}^2$$

$$A_r = 0.15 \text{ m}^2$$

$$A_c = 0.15 \text{ m}^2$$

Assuming the skin is completely wetted,

$$p_s = 47 \text{ mm Hg} \qquad T_w = T_a = 22^\circ\text{C}$$

$$p_a = 11 \text{ mm Hg} \qquad T_s = 35^\circ\text{C}$$

The initial rate of temperature drop is then calculated from (6.19) as

$$1.97\, dT/dt = 4.6 - 77.1 - 9.8 - 18.2 = -101.5 \text{ kcal/hr}$$

$$dT/dt = -51.5^\circ\text{C/hr} = -0.86^\circ\text{C/min} = -1.5^\circ\text{F/min}$$

This approximate calculation demonstrates why such immediate care must be taken during delivery procedures for quick reduction of thermal losses from newborns to preserve normal body temperatures.

While we have presented the elements of the applications of the general principles of heat transfer to prediction of the relationship of living systems and the environment, we should point out that in general this is a most complicated and detailed subject that is of considerable research interest. The physiological factors governing such parameters as sweat rates and skin temperatures continue to be a subject of much scientific inquiry. The purpose of this section is to indicate how such information, when available, can be used.

A summary of the overall energy balance equations and the forms of the equations available for individual contributions are presented in Table 6.3.

HEAT TRANSFER WITHIN THE MEDIUM

Inside living systems the significant modes of heat transfer are conduction and forced convection. The human body, for instance, is equipped with a multitude of concurrent and countercurrent heat

Table 6.3. A Summary of Environmental Exchange

$$\dot{M}_o + \Delta\dot{M} + \dot{T} = \hat{H}_f w_f - \dot{Q}_{resp} - \dot{Q}_{vap} - \dot{Q}_{rad} - \dot{Q}_{conv} - \dot{Q}_{cond} - \dot{W}$$

$\dot{M}_o$ = basal metabolism

$\Delta\dot{M}$ = excess metabolism $\quad \Delta\dot{M} = -n\dot{W} \quad n > 1$

$\dot{T} = m_{tot}\,\hat{C}_p\, dT/dt$

$\hat{H}_f$ = energy content of food, $= 4.1x_p + 4.1x_c + 9.3x_f$

w_f = flow rate at which food is supplied

$\dot{Q}_{resp} = \dot{V}/\widetilde{V}\,[(0.007(37 - T_I) + 10.8(0.066 - Y_I)]$

$\dot{Q}_{vap} = h_v\, A_v\, (p_s - p_a)$

$\dot{Q}_{rad} = \sigma\, e_w\, e_s\, A_r(T_s^4 - T_w^4) = h_r A_r (T_s - T_w)$

$\dot{Q}_{conv} = h_c\, A_c\, (T_s - T_a)$

$\dot{Q}_{cond} = k\,(T_s - T_w)/\Delta x$

$\dot{W}$ = physiological external work rate

exchangers. These elements apparently play a critical role in the regulation of organ and surface temperatures. In this regard there must be an inherent coupling effect between the internal and external heat exchange mechanisms.

To begin the specific discussion of internal heat exchange, it is necessary to define the following parameters.

Mean core temperature T_c. This quantity, usually measured in humans either orally, rectally, or at the tympanic membrane, is representative of the operating temperature of the internal organs. In humans, it is usually very close to 37°C (98.6°F).

Arterial blood temperature T_A. This temperature, which may vary throughout the system, represents the temperature of the fully oxygenated blood that perfuses a given organ or region of the body. It is usually close to the core temperature, although there is evidence that in some parts of the body this blood is precooled below core temperature by heat exchange, so that it can remove generated metabolic heat more effectively while simultaneously supplying oxygen.

Venous blood temperature T_v. This quantity, which also will vary over a significant range in a given location as well as between locations, represents the temperature of O_2-poor blood returning through the circulatory system. Since this blood usually passes closer to the surfaces than does arterial blood, it is more closely coupled to skin or surface temperature.

Surface temperature T_s. The surface or skin temperature varies regionally and with environmental conditions over a range of several degrees. Various weighting formulas are employed to represent the mean surface temperature.

Mean body temperature T_b. It is often desirable to obtain a weighted average temperature to characterize the temperature of the entire body. This takes into account skin and core temperature differences. Various weighting factors are used to compute this quantity, a typical set being

$$T_b = 0.6\ T_c + 0.4\ T_s \tag{6.20}$$

although various investigators recommend different sets, depending on the use for which T_b is intended.

Muscle temperature T_m. As demonstrated in Example 5.3, working muscle can experience significant temperature rise. It is therefore useful sometimes to take this into account in calculating the mean body temperature, for example

$$T_b = 0.15\ T_s + 0.30\ T_m + 0.55\ T_c \tag{6.21}$$

Subcutaneous fat thickness Δx_f. This quantity, which is on the order of 5 mm for adult humans, is indicative of the insulative capability of the system. The thermal conductivity of this substance (about 0.4 kcal/m hr °C) is such that the heat flux near the skin is substantially influenced by variations in its thickness.

Local blood flow rate w_{bi}. This quantity is often needed to estimate the convective heat transfer rate in a given region. Usually, the bloodstream is responsible for the major amount of heat transfer within the body. The local blood flow rate in the extremities changes dramatically depending on environmental conditions and mean body temperature. The amount and the distribution of this quantity are both very significant in evaluating the heat transfer situation.

The complexity of the w_{bi} situation is illustrated in Figure 6.2,

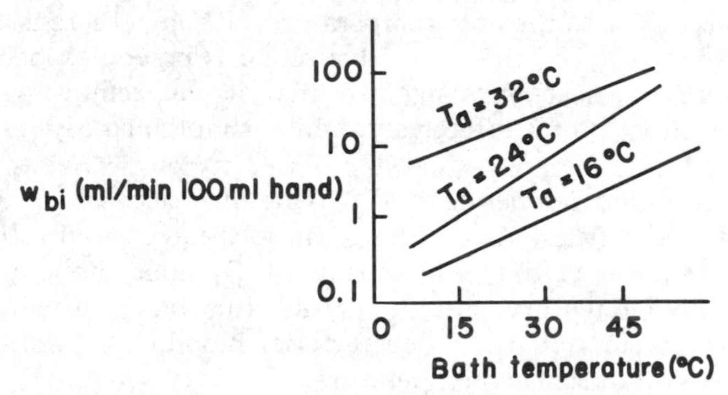

Fig. 6.2. Temperature effects on blood flow in the hand. (From C. R. Spealman, *Am. J. Physiol.* 145 [1945], 218.)

which gives some typical data for this quantity. This figure indicates the variation of the blood flow in the human hand as a function of the water bath temperature in which the hand is immersed. Also included is the effect of the room temperature in which the subject was seated, which is seen to be quite significant.

If we consider that the two major mechanisms for heat transfer inside the body are conduction through the body tissues and fluids and forced convection through the circulatory system, we may write quantitative expressions involving the variables defined above to describe some of the various situations that occur. A series of examples will be offered to illustrate this.

Example 6.8. Heat Losses in the Hand

Consider the hand as being perfused by arterial blood supplied from the body core and entering the hand at a temperature T_A. The venous blood returning to the arm will have a temperature T_v. The heat lost from the blood in the hand will then be

$$\dot{Q} = w_b \hat{C}_p (T_A - T_v) \tag{6.22}$$

At steady state this will be exactly equal to the heat lost from the hand to the surroundings if we neglect conductive gains or losses through the wrist. This is expressed as

$$\dot{Q} = h A (T_s - T_a) \tag{6.23}$$

where

h = overall heat transfer coefficient for the hand
A = area of the hand subject to external heat transfer

When the ambient air temperature is severely lowered, the organism reacts to reduce the heat lost from the hand, by lowering w_b and subsequently T_s in that region. Figure 6.3 indicates schematically

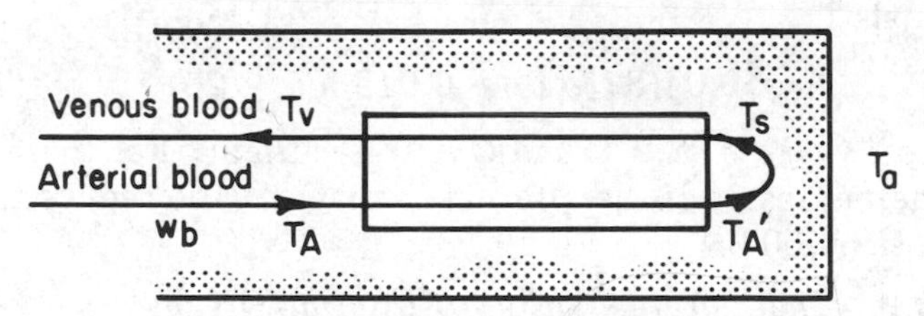

Fig. 6.3. Internal countercurrent heat exchange.

how the use of an internal countercurrent exchanger can further minimize the heat loss. The venous blood returning can cool the arterial blood being supplied to the extremity (or external heat exchanger) while it in turn (the venous blood) is itself being rewarmed by the arterial blood. At steady state,

$$w_b \hat{C}_p (T_A - T'_A) = w_b \hat{C}_p (T_v - T_s) \tag{6.24}$$

where T'_A represents the temperature of the precooled arterial blood being supplied to the skin surfaces, as depicted schematically in Figure 6.3. Close inspection of Figure 6.3 also reveals that

$$w_b \hat{C}_p (T'_A - T_s) = h A (T_s - T_a) \tag{6.25}$$

which is actually a combination of (6.22), (6.23), and (6.24).

For fixed T_A, T_a, h and A there are four independent variables left to control the total heat loss: w_b, T'_A, T_s and T_v. We have two independent equations available, and data is available on the variation of w_b with T_a. We can determine two of the variables as functions of the other two in order to demonstrate how the total loss can be minimized. Suppose that the following situation is set:

$T_a = 20°\text{C}$ $\qquad$ $A = 0.05\ \text{m}^2$

$T_A = 37°\text{C}$ $\qquad$ $\hat{C}_p = 1.0\ \text{kcal/1000 ml °C}$

$h = 15\ \text{kcal/m}^2\ \text{hr °C}$ $\qquad$ $w_b = 40\ \text{ml/min}$

Equations (6.22), (6.23), and (6.24) reduce to

$$-\dot{Q} = 0.04(37 - T_v) = 0.0125(T_s - 20) = 0.04(T'_a - Ts) \tag{6.26}$$

Now the selection of any one temperature will set the other two. For example, if $T_v = 35°\text{C}$, then $T_s = 26.4°\text{C}$ and $T'_A = 28.4°\text{C}$, and the rate of heat loss is 80 cal/min. The reader may determine other combinations. It is evident from these values that the effect of the internal heat exchanger is to reduce the rate of heat loss by lowering the surface temperature. This may be demonstrated most simply by reconsidering the problem without the internal exchanger. Then, if the surface were perfused with the same flow rate of warm blood, since T_s now equals T_v,

$$\dot{Q} = 0.04(37 - T_s) = 0.0125(T_s - 20) \tag{6.27}$$

which gives $T_s = T_v = 32.9°\text{C}$ and $\dot{Q} = 164$ cal/min. For these combinations, the net result of the internal exchanger is to reduce the total heat loss rate by 50%.

Example 6.9. Thermal Regulation of Internal Organs

There is considerable evidence that certain physiological systems also employ the combination of an internal countercurrent exchanger and an external exchanger to regulate the temperature of heat-generating organs such as the brain. Recent work has indicated that, in this case, cooled venous blood from the external exchanger may be acting to precool arterial blood leading to the organ by means of an internal exchanger. The precooled blood then not only supplies nutrients and removes waste products but also removes the excess heat from the organ. Figure 6.4 indicates one possible way that this

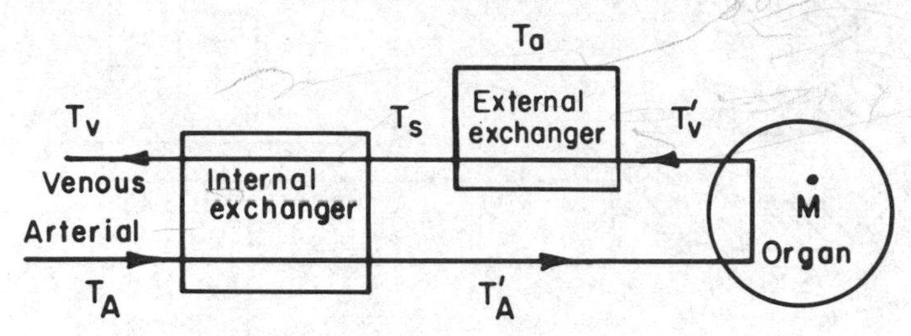

Fig. 6.4. Thermal regulation of organs.

might be accomplished by a continuous single stream.

As in the previous example, three energy balances may be written. Referring to the figure:

Internal exchanger:

$$w_b \hat{C}_p (T_A - T'_A) = w_b \hat{C}_p (T_s - T_v) \tag{6.28}$$

Organ:

$$w_b \hat{C}_p (T'_v - T'_A) = \dot{M} \tag{6.29}$$

External exchanger:

$$w_b \hat{C}_p (T'_v - T_s) = h A (T_s - T_a) \tag{6.30}$$

These equations can be used in developing an understanding of the thermal regulatory aspects of the above depicted flow scheme and can aid in interpreting experimental data obtained from such systems. It should be noted, however, that such systems and their mechanisms, which govern the distribution of blood flow, are quite complex. A better overall picture than either Figure 6.3 or 6.4 might be to depict a network of flows and exchangers. Nevertheless, these past two examples can be useful in elucidating how the quantitative treatment of heat transfer can aid in understanding the function of such systems as well as describing, if only in a limited way, their salient features.

Example 6.10. Internal Conduction through Composite Layers

In this example the relations for the steady temperature distribution through composite layers of core, muscle, and skin will be developed. Figure 6.5 depicts a simplified model of such a distribution. The cylindrical set of composite layers is sectioned and since the layers are thin compared to their diameter, the rectangular coordinate Δz's are used to measure the heat conduction paths and to

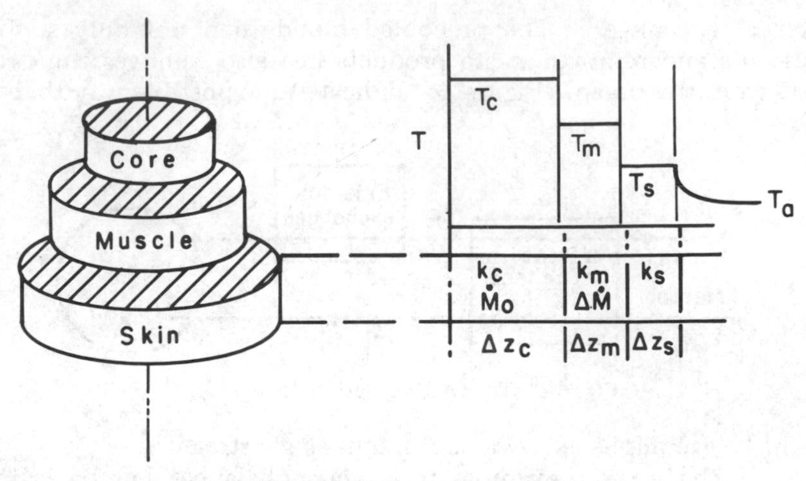

Fig. 6.5. Conduction model of the body.

describe the heat flux. At steady state the heat generated in the core layer must be conducted through each layer and ultimately transferred to the environment. If T_c, T_m, and T_s represent the average temperatures in each layer and if the thermal conductivities of each layer are equal, the heat fluxes may be expressed as

$$-\dot{M}_o = k(T_c - T_m)/\Delta z_{cm} = k(T_m - T_s)/\Delta z_{ms} = h(T_s - T_a) \qquad (6.31a)$$

heat generated in the core = heat conducted from core to muscle
= heat conducted from muscle to skin
= heat transferred to environment

where

$$\Delta z_{cm} = 0.5(\Delta z_c + \Delta z_m) \qquad (6.31b)$$

$$\Delta z_{ms} = 0.5(\Delta z_m + \Delta z_s) \qquad (6.31c)$$

For an $\dot{M}_o$ of -37 kcal/m² hr, an h of 10 kcal/m² hr °C, an ambient temperature T_a of 30° C, a core temperature of 37° C, and a thermal conductivity of 0.40 kcal/m hr °C, the skin and muscle layer temperatures will be determined for this model by the choice of layer thicknesses. Since there are three independent equations in (6.31a–b), one of the thicknesses will be determined by choosing the other two. If all the above variables are specified, one reasonable set of typical thicknesses for skin and muscle thicknesses would be $\Delta z_s = 1$ cm, $\Delta z_m = 2$ cm, for which, as the reader may confirm, $T_s = 33.3$, $T_m = 34.7$, and $\Delta z_c = 4.7$ cm. This gives a total thickness of the model as 15.4 cm, which would be a reasonable diameter for a limb.

Note that in this example it was convenient to deal with fluxes rather than flows, since no area dimensions were available or set by the model. Also note that the neglecting of the curvature appears justified because the total diameter was an order of magnitude larger than the layer thicknesses.

The excess metabolic energy production rate, $\Delta \dot{M}$, could be incorporated in this model by letting it be a generation term in the muscle layer, and adjusting (6.31a–b) accordingly.

Example 6.11. Simultaneous Convection and Conduction

Example 6.10 also provides a good medium for considering the effects of convective heat transfer. Suppose we superimpose a bloodstream on Figure 6.5, so that the situation is as shown in Figure 6.6,

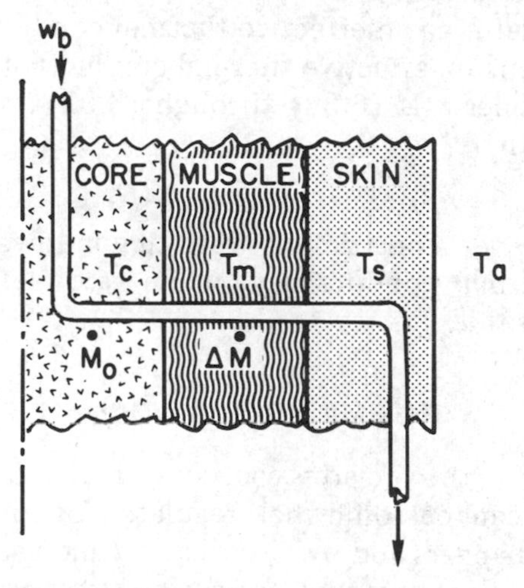

Fig. 6.6. Convective and conductive heat transfer.

albeit somewhat idealized. In addition to the heat transferred by conduction, additional heat will be carried through the layers by the blood flow. If we assume that the bloodstream approaches the temperature of each layer before it leaves (physically impossible in the limit as seen in Chapter 5), (6.31a–b) becomes

$$\begin{aligned} -\dot{M}_o &= k\,A\,(T_c - T_m)/\Delta z_{cm} + w_b\,\hat{C}_p\,(T_c - T_m) \\ &= k\,A\,(T_m - T_s)/\Delta z_{ms} + w_b\,\hat{C}_p\,(T_m - T_s) \\ &= h\,A\,(T_s - T_a) \end{aligned} \tag{6.32}$$

or, on combining;

$$\begin{aligned} -\dot{M}_0 &= (k\,A/\Delta z_{cm} + w_b\,\hat{C}_p)\,(T_c - T_m) \\ &= (k\,A/\Delta z_{ms} + w_b\,\hat{C}_p)\,(T_m - T_s) \\ &= h\,A\,(T_s - T_a) \end{aligned} \tag{6.33}$$

If heat is also generated in the muscle layer by an excess amount $\Delta\dot{M}$, then the steady-state equation may be adjusted to take this into account as in the previous example. The new steady-state equations are then

$$-\dot{M}_o = (k\,A/\Delta z_{cm} + w_b\,\hat{C}_p)\,(T_c - T_m) \tag{6.34}$$

$$-\Delta\dot{M} = (k\,A/\Delta z_{ms} + w_b\,\hat{C}_p)\,(T_m - T_s) - (k\,A/\Delta z_{cm} + w_b\,\hat{C}_p)\,(T_c - T_m) \tag{6.35}$$

$$(k\,A/\Delta z_{ms} + w_b\,\hat{C}_p)\,(T_m - T_s) = h\,A\,(T_s - T_a) \tag{6.36}$$

These relations indicate why many investigators choose to account for variable blood flows caused by vasodilation and vasoconstriction by defining an effective thermal conductivity. It is clear why these apparent or effective thermal conductivities will be higher than the actual ones. If (6.34) through (6.36) are written in conduction form, then

$$k_{\text{eff}} = k + w_b\,\hat{C}_p\,\Delta z/A \tag{6.37}$$

which shows how vasoconstriction (resulting in lowering w_b) reduces the effective thermal conductivity, while vasodilation (resulting in increased w_b) raises it.

THERMAL REGULATION

The examples in the previous section have demonstrated how surface heat losses can be somewhat regulated by the functioning of internal heat exchangers and by vasodilation and vasoconstriction, all of which result in the adjustment of surface temperature. Two other modes of temperature control mechanisms have been referred to previously and are appropriately reviewed here in the present context. Examination of Table 6.3 shows that the expression for body temperature changes for periods between meals and may be written as

$$\dot{T} = -(\dot{M}_o + \Delta\dot{M}) - \dot{Q}_{\text{resp}} - \dot{Q}_{\text{evap}} - \dot{Q}_{\text{rad}} - \dot{Q}_{\text{conv}} - \dot{Q}_{\text{cond}} - \dot{W} \tag{6.38}$$

The regulation of surface temperature affects the terms $\dot{Q}_{\text{rad}}$, $\dot{Q}_{\text{conv}}$, and $\dot{Q}_{\text{cond}}$; $\dot{M}_o$ by definition is set, and $\dot{W}$ is explicitly related to $\Delta\dot{M}$. Therefore, any additional mechanisms for temperature control must involve the terms $\Delta\dot{M}$, $\dot{Q}_{\text{resp}}$, or $\dot{Q}_{\text{evap}}$. This is of course exactly the case.

Increased metabolic conversion in the muscles, as in shivering, represents an increase in the absolute value of $\Delta\dot{M}$, and hence a stimulus for increasing or positive $\dot{T}$. Since the efficiency of shivering muscles is low, about 10–15%, considerable energy is converted to heat. Example 6.11 was instructive in explaining how this will affect the mean body temperature T_B. Other forms of physical exercise that use the muscles will have similar effects in raising T_B, unless thermal losses from the surface continue to predominate.

An increase in $\dot{Q}_{evap}$ can be effected by secretion of water to the skin surface, that is, sweating, provided that the water is evaporated. If the water runs off without evaporating, as in heavy sweating, then no cooling effect is realized from the loss of that water. The surface heat losses resulting from evaporation may be predicted with the use of such relations as (6.16) and may sometimes be large, as in Example 6.7; but as pointed out there, the area for evaporation and the partial pressure of water at the surface are difficult parameters to estimate. Both of these are evidently controlled to some degree by the sweating reflex. There appears to be considerable evidence that in both shivering and sweating the stimulus for increased activity is roughly proportional to the deviation from normal of the mean body temperature.

In many species, such as the dog, evaporative losses included in the term $\dot{Q}_{resp}$ may be used as thermal regulatory devices A panting dog loses heat from his wet tongue and tends to reduce his $\dot{T}$ accordingly. While this is rigorously an evaporative loss, it is so closely related to the respiratory rate that it is usually included there.

All four of the major thermal regulatory devices (vasodilation, vasoconstriction, shivering, and sweating) operate on the negative feedback principle so common in physiological systems. That is, a positive body temperature increase stimulates increased sweating and vasodilation, which in turn tends to increase surface heat loss, in the first instance by increasing evaporative loss and in the second by increased internal convection to the surface resulting in increased surface temperature, which in turn results in increased convective loss. The resulting increase in the cooling rate then feeds back and drives the original positive stimulus back toward zero. Approximate steady-state expressions to describe these effects can be written as

Effective conductivity change:

$$k_{eff}/k_{eff}^0 = 1 + a_k \, \Delta T_B \tag{6.39}$$

Sweating loss:

$$\dot{Q}_{evap}/\dot{Q}_{evap}^0 = 1 + g_e \, \Delta T_B \tag{6.40}$$

Shivering loss:

$$\Delta \dot{M} = \Delta \dot{M}^0 - a_m \, \Delta T_B \tag{6.41}$$

The superscript ^{0} refers to the variable when $\Delta T_B = 0$. These equations, along with those in Table 6.3 and the material in Examples 6.10 and 6.11, provide the basis for a description of thermal regulation in many living systems

SUMMARY

Applications of the fundamentals of heat transfer—that is, the transfer of thermal energy resulting from temperature differences—in physiology and medicine have been presented here. The engineering approach of conceptually modeling in a simplified fashion the salient features of very complicated systems has been continually used throughout. Of course, nature is more complex and exasperating than these oversimplified models would indicate. Nevertheless, this chapter and its contents most assuredly represent a logical and consistent starting point, not only in developing an understanding of such systems but also in their analysis. In addition, these relations as discussed can be of considerable use in the design and interpretation of future experiments.

Bibliography

The following references provide not only fundamental data but in addition excellent examples of application of the principles and techniques of this chapter:

Edholm, O. G., and Bacharach, A. L. *Physiology of Human Survival.* Academic Press, 1965.

Herzfield, C. M. *Temperature—Its Measurement and Control in Science and Industry*, vol. 3, part 3, (J. D. Hardy, ed.). Reinhold, 1963.

Ruch, T. C., and Patton, H. D. *Physiology and Biophysics.* Saunders, 1965.

The following works are typical of excellent sources for experimental results:

Colin, J., and Houdas, Y. Experimental determination of coefficients of heat exchange by convection of the human body. *J. of Appl. Physiol.* 22, no. 1(1967), 31–38.

Hardy, J. D., and Stolwijk, J. A. J. Partitional calorimetric studies of man during exposure to thermal transients. *J. of Appl. Physiol.* 21, no. 6(1961), 1799-1806.

An integrated simulation of processes discussed in Chapters 4 and 6 may be found in:

Seagrave, R. C., and Burkhart, L. E. Simulation of temperature, heat dissipation, exercise, and weight losses in man. *Analog/Hybrid Computer Educational Users Group Transactions* no. 3(1970), 55–69.

The material in Example 6.9 was developed from work currently being done in the biomedical engineering program at Iowa State University, under the direction of Dr. James Magilton and Dr. Curran Swift.

chapter 7

Fundamentals of Mass Transfer

INTRODUCTION

The transfer of particular chemical species from one phase to another and the movement of such species along gradients within phases are fundamentally important processes in any system involving more than one chemical component. The various driving forces—such as free energy, pressure, electrical charge, temperature, and concentration—that promote the transfer of such species are fundamental physical quantities of importance not only in living systems but also in a host of chemical and physical systems encountered in the practice of engineering. The transport of chemical species as the result of such driving forces is commonly called *mass transfer*. This chapter, which in many respects is analogous to Chapter 5, deals with the principles and methods involved in the treatment of mass transfer.

In many respects the physical principles that describe the transfer of mass are similar to those encountered in heat transfer. The situation is usually much more complex in mass transfer, however, since instead of having one major dependent variable (temperature), there are as many dependent variables as there are species present. In addition, whereas usually only one mode of heat transfer has to be considered at one time within a phase, mass transfer almost always involves at least two fundamental mechanisms simultaneously. It is therefore necessary to set out very carefully some basic definitions to begin the treatment of this subject.

CONCENTRATIONS, VELOCITIES, AND FLUXES

It is first necessary to define a consistent set of mass and molar concentrations as done in Chapter 2. These are reviewed in Table 7.1.

It is also necessary to differentiate between certain velocities that occur in mass transfer situations. We often refer to the *hydrody-*

Table 7.1. Concentration Relations

	Mass	*Molar*
Volumetric concentration	ρ_i (mass i/vol)	C_i (moles i/vol.)
Total density	ρ (mass/vol)	C (moles/vol)
Fraction	ω_i	x_i, y_i

Identities

$\rho_i = \omega_i \rho$	$C_i = X_i C$
$\Sigma_i \rho_i = \rho \quad \Sigma_i \omega_i = 1$	$\Sigma_i C_i = C \quad \Sigma_i X_i = 1$
$\rho_i = M_i C_i \quad \rho = MC$	$M = \Sigma_i X_i M_i$
$X_i = \dfrac{\omega_i/M_i}{\Sigma_i \omega_i/M_i}$	$\omega_i = \dfrac{X_i M_i}{\Sigma_i X_i M_i} = \dfrac{X_i M_i}{M}$

namic velocity v, which is a measure of the velocity of the bulk fluid stream relative to some fixed location. This is perhaps the most familiar velocity, and corresponds, for example, to what one would measure with a pitot tube or other velocity meter. It is necessary to also define a *species velocity* v_i, which represents the velocity of the species i relative to fixed coordinates, so that the relation between these two velocities is given by

$$v = \sum_i \omega_i v_i \tag{7.1}$$

that is, the hydrodynamic velocity v is really the *mass average velocity* of the fluid or system—a weighted average of all the species velocities. Obviously, for the flow of a single species the two are identical.

In a similar fashion we can define the *molar average velocity* v^* as

$$v^* = \sum_i x_i v_i \tag{7.2}$$

In general, the mass average velocity and the molar average velocity are not the same and are not equal, since only in the case of all of the species in the mixture having the same molecular weights would the mass fractions and mole fractions be equal.

With these basic reference velocities, it is now possible to define the *diffusion velocity* of a species i, v_{di}, as follows

$$v_{di} = v_i - v \tag{7.3}$$

$$v_{di}^* = v_i - v^* \tag{7.4}$$

The essence of developing a basic understanding of mass transfer lies in the notion, suggested by the definition of the diffusion velocity, that diffusion or molecular mixing of a species i occurs relative to the average velocity of the medium. In other words, diffusion is almost always superimposed on convective mass transfer. To illustrate this point further, we will define some fluxes.

Mass flux, n_i: The flux of species i relative to fixed coordinates, mass of i/area time

$$n_i = \rho_i v_i \tag{7.5}$$

Mass flux j_i: The flux of species i relative to the mass average velocity, mass of i/area time

$$j_i = \rho_i (v_i - v) \tag{7.6}$$

According to the definitions in Table 7.1 and (7.1) we see that

$$\sum_i n_i = \sum_i \rho_i v_i = \rho \sum_i \omega_i v_i = \rho v \tag{7.7}$$

and

$$\sum_i j_i = \sum_i \rho_i (v_i - v) = 0 \tag{7.8}$$

and, most importantly,

$$n_i = \omega_i \sum_i n_i + j_i \tag{7.9}$$

a result which the reader is urged to work out in detail.

Equation 7.9 is the single most important equation in the study of mass transfer. It says that the flux of a species i is the sum of the flux of i due to the bulk motion or convection of the stream, and the flux of i due to diffusion of i relative to the bulk average velocity.

For a binary (two-component) system, (7.9) is

$$n_A = \omega_A (n_A + n_B) + j_A = \omega_A \rho v + j_A \tag{7.10}$$

The molar equivalents of (7.5) through (7.10) are

Molar fluxes:

$$N_i = C_i v_i \tag{7.11}$$

$$J_i = C_i (v_i - v) \tag{7.12}$$

Identities:

$$\sum_i N_i = \sum_i C_i v_i = C v^* \tag{7.13}$$

$$\sum_i J_i = \sum_i C_i (v_i - v) = C(v^* - v) \tag{7.14}$$

$$N_i = x_i \sum_i N_i + J_i \tag{7.15}$$

$$N_A = x_A (N_A + N_B) + J_A \tag{7.16}$$

The fluxes j_A and J_A are the total mass transfer diffusional fluxes relative to the mass average velocity and are actually the sum of several fluxes resulting from different driving forces such as gradients in temperature, pressure, external forces, and concentration. For the time being we will restrict ourselves strictly to diffusion resulting from concentration gradients, not only because it will promote better understanding of the principles of mass transfer but also because in the majority of problems it is usually the most important phenomenon.

At this point it is then appropriate to discuss in detail the diffusion fluxes j_i and J_i. These fluxes may be expressed in terms of the concentration gradients, using the basic diffusion law presented by Fick. In rectangular coordinates,

$$j_i = - \rho \, \mathcal{D}_{im} \, d\omega_i/dz \tag{7.17}$$

$$J_i = - C \, \mathcal{D}_{im} \, dx_i/dz \tag{7.18}$$

$\mathcal{D}_{im}$ is the diffusion coefficient for species i in the medium being considered. We will discuss this quantity in detail subsequent to an examination of the physical forms of (7.17) and (7.18).

The reader should first note the similarity of form between these two expressions and Fourier's law for heat conduction. In both cases the form is rate = conductance × driving force.

For such media as dilute solutions where the total density and total concentration are nearly constant with position (that is, homogeneous), the flux expressions may be written as

$$j_i = - \mathcal{D}_{im} \, d\rho_i/dz \tag{7.19}$$

$$J_i = - \mathcal{D}_{im} \, dC_i/dz \tag{7.20}$$

In cylindrical and spherical coordinates, the radial fluxes may be written by replacing dz with dr.

The quantity $\mathcal{D}_{im}$ is on the order of 10^{-5} cm^2/sec for liquids and 10^{-1} cm^2/sec for gaseous systems. For gases it is proportional to the 3/2 power of absolute temperature and is relatively independent of pressure. For liquids it is roughly linear with temperature. Table 7.2 gives some typical values for $\mathcal{D}_{im} = \mathcal{D}_{AB}$ in binary mixtures.

Table 7.2. Typical Values of Diffusion Coefficients

Gases			Liquids		
System	*Temp*	$\mathfrak{D}_{AB}$*	*System*	*Temp*	$\mathfrak{D}_{AB}$
CO_2-N_2	0°C	0.144	Ethanol-H_2O	25°C	1.13×10^{-5}
	25°C	0.165	CO_2-H_2O	10	1.46
H_2O-air	25	0.258		20	1.77
	60	0.305	HCI-H_2O	0	1.80
O_2-N_2	0	0.181		10	2.50

*cm^2/sec

Unfortunately, the diffusion coefficient is often a strong function of concentration. Since it is a parameter which describes the random mixing of molecules of different species in regions where concentration differences exist, it is not surprising that it should be concentration dependent. Since many problems of interest in living systems involve very dilute solutions where the concentration difference is weak, this feature will not be too troublesome for our purposes.

STEADY-STATE DIFFUSION

For single-phase systems with no chemical reactions occurring, the steady-state flows in and out of any differential element of the phase must be equal, and one-dimensional equations applicable at any point in the system may be written as follows:

Rectangular geometry:

$$N_i = \text{constant} = x_i \sum_i N_i - \mathfrak{D}_{im} \, dC_i/dz \tag{7.21}$$

Cylindrical geometry:

$$rN_i = \text{constant} = r \, x_i \sum_i N_i - r \, \mathfrak{D}_{im} \, dC_i/dr \tag{7.22}$$

Spherical geometry:

$$r^2 N_i = \text{constant} = r^2 \, x_i \sum_i N_i - r^2 \, \mathfrak{D}_{im} \, dC_i/dr \tag{7.23}$$

The constants in these expressions are usually evaluated at the boundaries of the system, using an appropriate physical boundary condition.

Example 7.1 Evaporation through a Stagnant Film

To illustrate the use of the steady-state expressions given above, we shall consider the evaporation of water from a surface, under the condition where there is a relatively stagnant film of air near the surface. Letting the subscript A refer to water and B to air, then the fluxes are

$$N_B = 0 \text{ (the air is stagnant)} \tag{7.24}$$

$$N_A = x_A N_A - C \mathfrak{D}_{AB} \, dx_A / dz \tag{7.25}$$

or

$$N_A = (-C \mathfrak{D}_{AB})/(1 - x_A) \frac{dx_A}{dz} \tag{7.26}$$

Equation 7.25 is a statement that the flux of water away from the surface is the sum of that carried away by the motion of the water itself, pushing its way through the air and resulting in an overall flow, and that which diffuses relative to the overall flow because of a concentration difference.

Equation 7.21 now becomes

$$N_A = \frac{(-C \mathfrak{D}_{AB})}{(1 - x_A)} \frac{dx_A}{dz} = \text{a constant} \tag{7.27}$$

Since the water is evaporating at the surface, the mole fraction at that point is known. At $z = 0$, $x_A = x_{AS} = p_{AS}/P$, where p_{AS} = the vapor pressure of water at the surface temperature. Further, a good approximation is that at a distance Δz from the surface, the concentration of water is equal to the ambient concentration; that is, at $z = \Delta z$, $x_A = x_{A\infty} = 0$ in this problem.

With these two boundary conditions (7,27) can be readily integrated and the constant evaluated to give first the concentration profile,

$$\frac{1 - x_A}{1 - x_{AS}} = \left(\frac{1}{1 - x_{AS}} \right)^{z/\Delta z} \tag{7.28}$$

and the flux at any point—the constant in (7.27);

$$N_A = \frac{C \mathfrak{D}_{AB}}{\Delta z} \ln (1/1 - x_{AS}) \tag{7.29}$$

which in terms of partial pressures may be written as

$$N_A = \frac{P \mathfrak{D}_{AB}}{R \, T \Delta Z} \ln (P/P - p_{AS}) \tag{7.30}$$

where P is the total pressure in the system, r is the gas constant, and T is the temperature.

If the convection term (the bulk flow term) in (7.25) is neglected, (7.26) becomes simply

$$N_A = -C\,\mathfrak{D}_{AB}\, dx_A/dz \tag{7.31}$$

and, as the reader may confirm, the flux equivalent to (7.30) is

$$N_A = (\mathfrak{D}_{AB}/RT)(p_{AS}/\Delta_z) \tag{7.32}$$

We recall from Chapter 6 that the heat flux from the surface due to evaporation may be written as

$$\dot{Q}_{\text{evap}} = h_v\, A_v\, (p_{AS} - p_{A\infty}) \tag{7.33}$$

For this case, where $p_{A\infty} = 0$,

$$\dot{Q}_{\text{evap}} = h_v\, A_v\, p_{AS} \tag{7.34}$$

Now, the heat removed from the surface can also be written as the product of the mass flux, the area, and the heat of vaporization (that is, the enthalpy change resulting from vaporization) as follows:

$$\dot{Q}_{\text{evap}} = N_A\, A_v\, \Delta\widetilde{H}_v \tag{7.35}$$

If this expression is combined with (7.32), the basic relation for the evaporative heat transfer coefficient is given by

$$h_v = (\mathfrak{D}_{AB}\, \Delta\widetilde{H}_v)/(RT\, \Delta z) \tag{7.36}$$

This result implies that the evaporative heat transfer coefficient is a strong function of surface temperature.

Example 7.2 Diffusion from the Surface of a Spherical Drop

In this example we will consider the steady-state evaporation of a volatile liquid from the surface of a spherical drop. Using (7.23), and the stagnant film approximation; $N_B = 0$,

$$N_A = \frac{-C\,\mathfrak{D}_{AB}}{1 - x_A}\frac{dx_A}{dr} = \frac{C}{r^2} \tag{7.37}$$

with the boundary conditions at the surface, $x_A = x_{AS}$, and at a long distance, $x_A = 0$.

The concentration profile for this case then becomes

$$\ln(1 - x_A) = \ln(1 - x_{AS})\, R/r \tag{7.38}$$

and the surface flux is

$$N_{AS} = \frac{C\,\mathfrak{D}_{AB}\ln(1 - x_{AS})}{R} \tag{7.39}$$

where R is the drop radius.

Example 7.3 Diffusion into a Falling Liquid Film

In this example we will consider the absorption of a solute from a gas phase into a moving film of liquid. In dilute liquid solutions the convective term in the steady-state flux expression can often be safely neglected if there is no forced convection in the direction being considered. In this example, as depicted in Figure 7.1, there is a

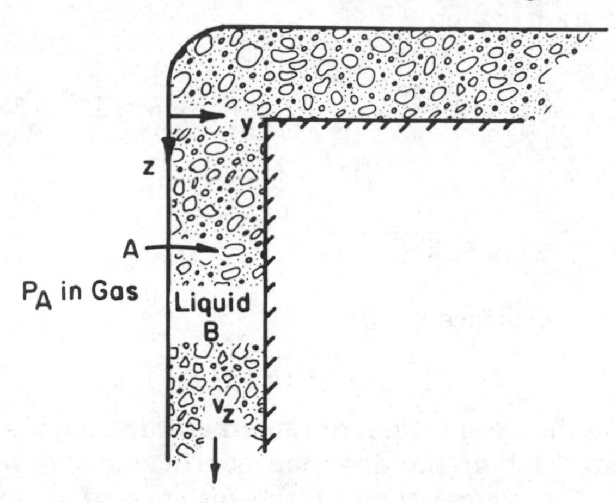

Fig. 7.1. Falling film with gas absorption.

diffusion flux in the y direction, which may be written as

$$N_{Ay} = - \mathfrak{D}_{AB}\ (\partial C_A / \partial y) \tag{7.40}$$

and a convective flux in the z direction, in which the convection of solute due to the bulk motion of the fluid is much more important than the diffusion in that direction:

$$N_{Az} = x_A\ (N_{Az} + N_{Bz}) = C_A\ v_z \tag{7.41}$$

An overall material balance on a differential element of fluid, using the two fluxes given above, leads to a two-dimensional equivalent of (7.21), a partial differential equation in y and z;

$$\mathfrak{D}_{AB}\ (\partial^2 C_A / \partial y^2) = v_z\ (\partial C_A / \partial z) \tag{7.42}$$

A typical set of boundary conditions for this situation might be

1. At $z = 0$, $C_A = 0$ (no absorption yet).
2. At $y = 0$, $C_A = C_{AS}$ (the surface value).
3. At $y = \infty$, $C_A = 0$ (the solute does not "penetrate" very far into the liquid, so that for practical purposes this is an appropriate condition).

The relation between C_{AS} and p_{AS} for this situation, and for many similar gas-liquid mass transfer problems, may be given by an expression such as Henry's law, which is an equilibrium relationship between gas phase partial pressure and liquid phase concentration.

$$p_{AS} = H\ C_{AS} \tag{7.43}$$

where H = Henry's law constant. This linear relationship is often satisfactory for systems where the solution is sufficiently dilute.

The solution for this particular boundary value problem with this equilibrium relation is then

$$C_A/C_{AS} = 1 - (2/\sqrt{\pi}) \int_0^{\eta} \exp(-\eta^2)d\eta = 1 - erf\,\eta \qquad (7.44)$$

where

$$\eta = y/\sqrt{4\ \mathfrak{D}_{AB}\ z/v_z}$$

The flux at the surface $y = 0$ is

$$N_{AS} = C_{AS}\sqrt{(\mathfrak{D}_{AB}\ v_z)/\pi\ z} \qquad (7.45)$$

In this case the surface flux, or rate of absorption, is seen to vary with the square root of the diffusion coefficient and velocity and inversely with the square root of the distance of exposure. The absorption rate decreases with distance since the fluid becomes more nearly saturated. The ratio z/v_z is equivalent to the exposure time.

The mathematical complexity involved in this problem is typical of forced convective mass transfer problems and provides good evidence for the convenience of mass transfer coefficients for convective mass transfer problems. A later section in this chapter will deal more extensively with this topic.

DIFFUSION AND CHEMICAL REACTION

When we consider the effect of chemical reaction involving the diffusing species on the mass transfer situation, we find it convenient to differentiate between two major classes of chemical reaction situations.

Homogeneous reactions take place exclusively within the phase being considered, while *heterogeneous reactions* take place at phase boundaries, interfaces, or surfaces. The method of treatment for these two classes is summarized and then illustrated by some examples.

In mass transfer problems involving homogeneous reactions, the reaction rate term that contains kinetic information and describes the rate of generation or destruction of the species in question is included in the material balance equation, similar to the way that heat generation terms are treated in Chapter 5. The steady-state material balance equations (7.21–7.23) then become

Rectangular geometry:

$$\frac{dN_i}{dz} = \frac{d}{dz}\left(x_i \sum_i N_i - \mathfrak{D}_{im}\frac{dC_i}{dz}\right) = -R_i \tag{7.46}$$

Cylindrical geometry:

$$\frac{1}{r}\frac{d}{dr}(r N_{ir}) = \frac{1}{r}\frac{d}{dr}\left(r\, x_i \sum_i N_i - r\,\mathfrak{D}_{im}\frac{dC_i}{dr}\right) = -R_i \tag{7.47}$$

Spherical geometry:

$$\frac{1}{r^2}\frac{d}{dr}(r^2 N_{ir}) = \frac{1}{r^2}\frac{d}{dr}\left(r^2 x_i \sum_i N_i - r^2\,\mathfrak{D}_{im}\frac{dC_i}{dr}\right) = -R_i \tag{7.48}$$

where R_i = the volumetric rate at which species i is produced by reaction. The nonreaction balances (7.21–7.23) are simply the integrated forms of these balances with R_i equal to zero.

Example 7.4 Absorption and Reaction in a Liquid

In this example we consider the case where a solute A reacts irreversibly with a liquid solvent B, so that $A + B \longrightarrow AB$. We assume further that the rate at which A reacts at any point in the solution is proportional to the concentration of A at that point, that is, a first-order reaction. The absorption of oxygen into a hemoglobin solution is a good example of this situation, as depicted in Figure 7.2.

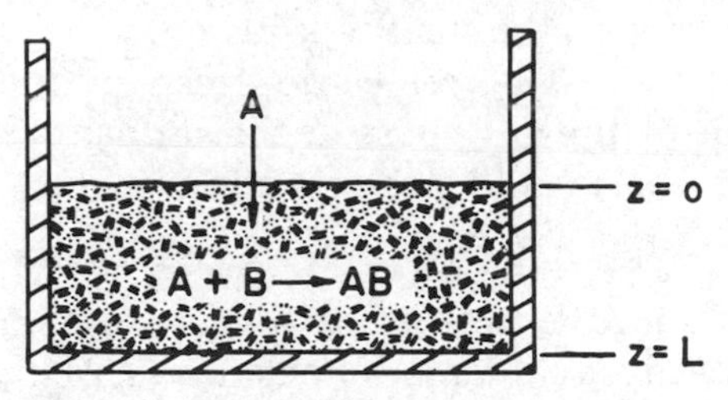

Fig. 7.2. Diffusion with homogeneous reaction.

If the concentration of A in the liquid is always low, then once again convection is not significant, and (7.46) becomes

$$\mathfrak{D}_{AB}\,(d^2C_A/dz^2) = -R_A = -KC_A \tag{7.49}$$

where K is the first-order reaction rate constant and has the units of (volume $\times$ time)$^{-1}$.

The boundary conditions for this problem are:

1. At $z = 0$, $C_A = C_{AS}$ ($C_{AS} = p_{AS}/H$).
2. At $z = L$ no mass transfer through the bottom can occur, so $dC_A/dz = 0$.

This condition is analogous to an insulated surface in heat transfer. The solution for the concentration profile is then

$$C_A/C_{AS} = \frac{\cosh\left[\left(\sqrt{KL^2/\mathfrak{D}_{AB}}\right)(1 - z/L)\right]}{\cosh\sqrt{KL^2/\mathfrak{D}_{AB}}} \tag{7.50}$$

where cosh = the hyperbolic cosine. The surface flux, or rate of absorption, is then given by

$$N_{AS} = -\ \mathfrak{D}_{AB}\ (dC_A/dz)_{z=0} = \sqrt{K\ \mathfrak{D}_{AB}}\ \ \tanh\sqrt{KL^2/\ \mathfrak{D}_{AB}} \tag{7.51}$$

where tanh = the hyperbolic tangent.

Note that in this case the flux varies throughout the liquid from a maximum at the surface to zero at the bottom.

In problems with heterogeneous reactions the reaction rate information is usually introduced as a boundary condition, since the reaction takes place at the surface or boundary of the phase. The following example is illustrative of this.

Example 7.5. Diffusion and Heterogeneous Chemical Reaction

Suppose that a solute species A in the gas phase, with a mole fraction x_A, is diffusing through a thin gas film near a solid plane surface and at that surface is reacting to form A_2, so that two molecules of A are dimerizing to form one molecule of A_2. Suppose also that the reaction rate at the surface is proportional to the concentration of A at the surface. Then, since no A can go through the surface, the flux of A out of the gas phase at the surface will be equal to the rate at which A is consumed by the reaction, that is,

$$N_{AS} = K\ C_A = C\ K\ x_A \quad \text{at } z = z_s \tag{7.52}$$

if z is measured from the outside of the gas film, and z_s is the film thickness. The steady-state material balance (7.21), then is

$$N_A = x_A\ (N_A + N_{A_2}) - C\ \mathfrak{D}_{AM}\ dx_A/dz = \text{a constant} \tag{7.53}$$

From the stoichiometry,

$$2\,N_{A_2} = -N_A \tag{7.54}$$

since every mole of the dimer uses up two moles of A. Then on substitution,

$$N_A = \frac{-C\ \mathfrak{D}_{AM}}{(1 - x_A/2)}\frac{dx_A}{dz} = \text{a constant} \tag{7.55}$$

with the boundary conditions

$$z = 0, \; x_A = x_{A0}$$
$$z = z_s, \; x_A = N_{AS}/CK = N_A/CK$$

Integrating and solving for the constant, the expression for the flux becomes

$$N_A = \frac{2C\,\mathfrak{D}_{AM}/z_s}{(1 + \mathfrak{D}_{AM}/Kz_s)} \ln\left(\frac{1}{1 - x_{A0}/2}\right) \tag{7.56}$$

which shows that the rate at which A is consumed at the surface is related to the diffusion coefficient of A in the mixture as well as the reaction rate constant and film thickness.

A "recipe" for solving mass transfer problems for steady-state situations can now be offered as follows:

1. Select the proper material balance equation (7.21, 7.22, 7.23 or 7.46, 7.47, 7.48).
2. Work out the relationships between the fluxes in order to simplify the material balance equation (example—stagnant film: $N_B = 0$).
3. If the media are dilute (x_A much less than 0.1), simplification by neglecting convective effects might be in order.
4. If chemical reaction is involved, decide on whether a homogeneous or heterogeneous model is more appropriate. Then select the appropriate reaction rate expression.
5. Choose appropriate boundary conditions.
6. Integrate and solve for the concentration profile and the flux expression. Sometimes the concentration profile is not required and does not have to be obtained to get the flux expression.

UNSTEADY-STATE DIFFUSION

For dilute media with no chemical reaction where convection is not significant, a form of Fick's law is often used and referred to as Fick's second law. In one-dimensional form it is written as

Rectangular geometry:

$$\frac{\partial C_A}{\partial t} = \mathfrak{D}_{AB} \frac{\partial^2 C_A}{\partial z^2} \tag{7.57}$$

Cylindrical geometry:

$$\frac{\partial C_A}{\partial t} = \mathfrak{D}_{AB} \frac{1}{r} \frac{\partial}{\partial r}\left(r \frac{\partial C_A}{\partial r}\right) \tag{7.58}$$

Spherical geometry:

$$\frac{\partial C_A}{\partial t} = \mathfrak{D}_{AB} \frac{1}{r^2} \frac{\partial}{\partial r}\left(r^2 \frac{\partial C_A}{\partial r}\right) \tag{7.59}$$

Because mathematical solutions of transient mass transfer problems are as complicated as the analogous heat transfer problems, there are few situations that can be solved directly without resort to numerical solutions or graphical methods. Very often, when the differential equations and boundary conditions are mathematically equivalent, results for heat transfer problems, such as in Example 5.2, can be used for mass transfer problems, as in the following example.

Example 7.6. Transient Diffusion in a Sphere

In this example we consider the leaching of a solute from a solid sphere being continually washed by a liquid solvent. If the original concentration of solute in the sphere is C_{A0} and the concentration in the solution is C_{AS}, (7.59) is the appropriate material balance equation and this problem is exactly analogous to the transient cooling of a solid sphere discussed in Example 5.2; the problem can therefore be solved with the use of Figure 5.2, replacing $\alpha\ t/R^2$ by $\mathfrak{D}_{AB}\ t/R^2$ and $(T - T_0)\ (T_1 - T_0)$ by the ratio $(C_A - C_{A0})\ (C_{AS} - C_{A0})$ where $\mathfrak{D}_{AB}$ is the diffusion coefficient of the solute in the sphere. Once again we must have very good stirring to maintain the surface concentration at a constant, C_{AS}. Diffusion coefficients in solids are usually on the order of 10^{-8} cm^2/sec or less, so for a 1 mm sphere, times on the order of 10^6 sec would be required to lower the concentration in the sphere at the center by measurable amounts (10^6 sec $\sim$ 10 days). If liquid drops in immiscible solvents were subjected to the same process, the time would be on the order of 10^3 sec (20 min) for the same concentration changes.

CONVECTIVE MASS TRANSFER

As in the case of convective heat transfer, exact solution of mass transfer problems where the primary mechanism is convection is almost never feasible. Resort to the concept of a mass transfer coefficient is necessary, as indicated by the analogous use of this type of coefficient for heat transfer problems. The situation for mass transfer will be somewhat more complex, however, for two reasons. First, the film thickness through which material diffuses will be affected by the rate of mass transfer, especially when the rate is high, so that the rate of transfer will affect the mass transfer coefficient itself. The second reason is that the flux expressions themselves are more complicated in mass transfer, since both diffusion and convection must be accounted for.

We will be most concerned with situations where the mass transfer rate is relatively low, so variations in the coefficients with rate will be unimportant.

A generalized mass transfer coefficient k_x may be defined as follows:

$$N_i = x_i \sum_i N_i + k_x \, \Delta x_i \qquad (7.60)$$

For dilute solutions, where x_i is very low,

$$N_i = J_i = k_x \, \Delta x_i = C \, \mathfrak{D}_{im} \, dx_i/dz \qquad (7.61)$$

and for transfer across a dilute film of thickness z_f,

$$k_x = C \, \mathfrak{D}_{im}/z_f \qquad (7.62)$$

The dimensionless physical quantity used to correlate mass transfer coefficients which is analogous to the Nusselt number (5.16) and represents an equivalent ratio of film thickness to diameter, is the *Sherwood number*, defined as

$$Sh = k_x \, D/C \, \mathfrak{D}_{im} \qquad (7.63)$$

It is common practice also to define mass transfer coefficients for gases and liquids using partial pressure and concentration driving forces respectively as

Gases:

$$N_i = k_g \, \Delta p_i \qquad (7.64)$$

Liquids:

$$N_i = k_l \, \Delta C_i \qquad (7.65)$$

The coefficients k_g and k_l may be related by Henry's law, as the reader may demonstrate.

The Sherwood number for many mass transfer situations may be correlated in terms of basic dimensionless quantities in a fashion similar to the Nusselt number. For forced convection problems, the Reynolds number reappears, while the Prandtl number is replaced by its analogous partner for mass transfer, the *Schmidt number*, defined as the ratio of the kinematic viscosity to the diffusivity, or

$$Sc = \mu/\rho \cdot \mathfrak{D}_{im} \qquad (7.66)$$

For free convective mass transfer problems the *mass transfer Grashof number* replaces its heat transfer analog, and is defined

$$Gr_{AB} = \rho^2 \, \mathfrak{L} \, g \, D^3 \, \Delta x_A/\mu^2 \qquad (7.67)$$

where $\mathfrak{L}$ = the concentration coefficient of volumetric expansion, and Δx_A is the characteristic mole fraction difference.

To convert any heat transfer correlation into the analogous or corresponding mass transfer correlation, it is often only necessary to substitute the proper mass transfer dimensionless groups. For example, the correlation for forced convection around a sphere, which for heat transfer is

$$Nu = 2.0 + 0.60\, Re^{0.5}\, Pr^{0.33} \quad (7.68)$$

becomes, for mass transfer

$$Sh = 2.0 + 0.60\, Re^{0.5}\, Sc^{0.33} \quad (7.69)$$

Once the mass transfer coefficient, however it is defined, has been determined, the overall mass transfer rate W_i may be computed using the product of the flux N_i and the interfacial area for mass transfer S. In many cases, however—such as in sprays, packed columns, and lungs—the interfacial area is not only unknown but is difficult to measure, and it becomes convenient to redefine the mass transfer coefficient as

$$W_i = N_i\, S = k_x a\, \Delta x_i\, V \quad (7.70)$$

where $k_x a$ = a new mass transfer coefficient, with a representing the area for mass transfer per unit volume, and V = the system volume. Then the quantity $k_x a$ is correlated and measured, rather than k_x.

Example 7.7. Evaporation from a Falling Drop

In this example we will consider a drop of liquid falling at a constant velocity v. The ambient air has a temperature T_A and is free of the drop material. An energy balance around the drop yields

$$W_A\, \Delta\widetilde{H}_v = 4\, \pi\, R^2\, h\, (T_S - T_A) \quad (7.71)$$

and the molar flow is given by

$$W_A = k_x\, 4\, \pi\, R^2\, x_{AS} \quad (7.72)$$

providing that x_{AS}, equal to the ratio of the vapor pressure of the liquid to the ambient total pressure, is very much less than one.

Values for h and k_x are obtained for this situation from (7.68) and (7.69), evaluating the physical properties at the film temperature T_f, where T_f is evaluated as the arithmetic average of the surface and ambient temperatures. For example, for a 70° F water drop evaporating in air at 140° F, T_f = 105° F, Sc = 0.58, and Pr = 0.80.

In general, however, the surface temperature is unknown, and Equations (7.71) and (7.72) must be combined to form

$$k_x\, x_{AS}\, \Delta\widetilde{H}_v = h\, (T_S - T_A) \quad (7.73)$$

and solved along with the vapor pressure relation

$$x_{AS} = p_A\, (T_S)/P \quad (7.74)$$

where p_A is the vapor pressure of A at T_s. This is a good example of a problem for which easy solution requires iteration or trial and error, since the vapor pressure relation is usually most readily available in tabular or graphical form. In addition, the values of h and k_x must be corrected each time a new surface temperature is assumed.

A new interpretation for the evaporative heat transfer coefficient discussed in (7.33) through (7.36) is now possible. Using the appropriate form of the flux expression

$$\dot{Q}_{evap} = W_A \, \Delta\widetilde{H}_v = k_x \, A_v \, \Delta x_A \, \Delta\widetilde{H}_v = k_x \, A_v \, \Delta\widetilde{H}_v \, \Delta p_A/P \quad (7.75)$$

so that h_v from (7.33) now becomes

$$h_v = k_x \, \Delta\widetilde{H}_v/P \quad (7.76)$$

Since the mass transfer coefficient k_x has the units of moles/area time, the units of h_v check out as energy/area time pressure difference.

Example 7.8. Mass Transfer across a Gas-Liquid Interface

This final example illustrating the use of mass transfer coefficients will deal with a very common engineering and physiological situation—the transfer of a solute from the liquid phase to the gas phase. The original approach to this situation was developed by Whitman in 1923 and was a monumental contribution in the field of chemical engineering.

Figure 7.3 indicates the situation schematically. A solute A is being transferred from the gas phase to a nonvolatile liquid solvent.

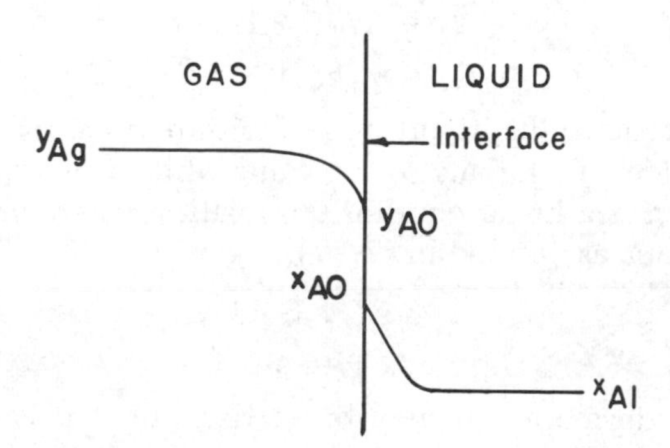

Fig. 7.3. Mass transfer at a gas-liquid interface.

Unlike previous examples such as Example 7.3, the resistances to transfer in both phases are important. At the interface there is a discontinuity of the mole fraction, since the solubility in the liquid is not necessarily the same as in the gas. The equilibrium relation that applies at the interface can be obtained from Henry's law or from some other solubility relationship. For example, from Henry's law,

$$y_{A0} = C\,H\,x_{A0}/P = m\,x_{A0} \tag{7.77}$$

where

y_{A0} = mole fraction in the gas phase at equilibrium at the interface

x_{A0} = mole fraction in the liquid phase at equilibrium at the interface

The mass transfer rate across the interface, again for dilute solutions, may be written as

$$N_A = k_y\,(y_{Ag} - y_{A0}) = k_x\,(x_{A0} - x_{Al}) \tag{7.78}$$

where k_y is the gas phase mass transfer coefficient.

Unfortunately, conditions at the interface are practically impossible to measure. It becomes necessary, as in the case of heat conduction through composite layers, to treat the situation without employing the interfacial conditions. The approach will be to use overall coefficients (as in the heat transfer case) by defining either a fictitious liquid side concentration or a fictitious gas side concentration as

$$N_A = K_x\,(x_A^* - x_{Al}) \tag{7.79}$$

or

$$N_A = K_y\,(y_{Ag} - y_A^*) \tag{7.80}$$

where the coefficients K_x and K_y are defined by

$$1/K_x = 1/k_x + 1/m\,k_y \tag{7.81}$$

$$1/K_y = m/k_x + 1/k_y \tag{7.82}$$

and x_A^* is the fictitious liquid mole fraction which is in equilibrium with y_{Ag}, and y_A^* is the analogous value which is in equilibrium with x_{Al}. That is, for a linear equilibrium relation where m is the equilibrium curve slope as defined in (7.77),

$$x_A^* = y_{Ag}/m \tag{7.83}$$

$$y_A^* = m\,x_{Al} \tag{7.84}$$

A similar development may be carried out if it is desired to use partial pressures in the gas phase or concentrations in the liquid phase, rather than mole fractions. Then,

$$N_A = K_G\,(p_{Ag} - p_{Al}) \tag{7.85}$$

where

$$1/K_G = H/k_l + 1/k_g \tag{7.86}$$

and

$$N_A = K_L\,(C_A^* - C_{Al}) \tag{7.87}$$

where

$$1/K_L = 1/k_l + 1/H k_g \tag{7.88}$$

where the p's and C's are defined analogously to the x's and y's above.

It should be pointed out that two fluid mass transfer problems frequently have unknown interfacial area and in addition often involve chemical reactions. This is certainly the case in many physiological applications. The basic principles presented in this example represent a starting point for the analysis of such situations.

Example 7.9. Mass Transfer across a Membrane

In this example we will consider the transfer of a solute from one liquid to another, with the two liquids separated by a membrane which is permeable to the solute. Figure 7.4 schematically depicts

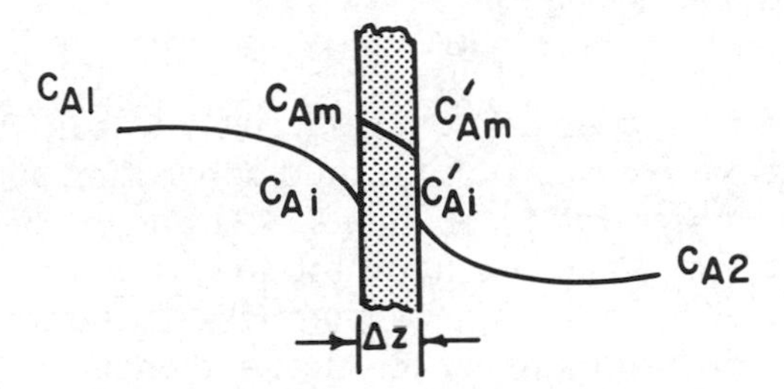

Fig. 7.4. Mass transfer through a membrane.

the situation. We will ignore any other mass transfer mechanisms than ordinary concentration-driven diffusion. In Chapter 8 we will take a more sophisticated look at this situation. This time, instead of mole fractions we will employ concentrations and once again restrict ourselves to dilute solutions.

The flux in terms of the concentration gradients shown in Figure 7.4 can be written as

$$N_A = k_1 (C_{A1} - C_{Ai}) = k_m (C_{Am} - C'_{Am}) = k_2 (C'_{Ai} - C_{A2}) \tag{7.89}$$

The concentrations in the membrane are related to the interfacial concentrations by

$$C_{Am}/C_{Ai} = C'_{Am}/C'_{Ai} = \phi \tag{7.90}$$

where ϕ is the distribution coefficient and is dependent on the solubility of the solute in the membrane and k_m is the diffusion coefficient of the solute in the membrane divided by the membrane thickness or the mass transfer coefficient for the membrane.

As in the previous example, the interfacial and membrane concentrations are not easily measurable, and in addition only the bulk solution concentrations are of primary interest. Therefore, an overall mass transfer coefficient is defined as

$$1/K = 1/k_1 + 1/\phi\, k_m + 1/k_2 \tag{7.91}$$

Since the quantity k_m is defined above, (7.91) may be written as

$$1/K = 1/k_1 + \Delta z/\phi\, \mathfrak{D}_{Am} + 1/k_2 \tag{7.92}$$

The k's in these two expressions are also referred to as *permeabilities.* The liquid phase mass transfer coefficients k_1 and k_2 will depend on the flow past the membrane as well as the properties of the fluids, as indicated earlier, while k_m is dependent on the properties and structure of the membrane. In Chapter 8, applications of this situation in hemodialysis will be discussed.

SUMMARY

This chapter has been devoted exclusively to a discussion of mass transfer by the mechanisms of bulk fluid convection and by diffusion as a result of concentration gradients. Although the principles of analysis remain the same, we have not specifically treated such important forms of mass transfer as thermal diffusion, pressure diffusion, or forced diffusion. Other physical and chemical processes that result in concentration differences—such as osmosis, filtration, and active transport—are also forms of mass transfer of importance in living systems, and they will be discussed and illustrated in chapter 8.

Bibliography

While the basic principles and working equations of mass transfer have been set out here, many readers undoubtedly desire more advanced and possibly more sophisticated treatment of many of these topics. The following references are recommended for such pursuits:

Bird, R. B., Stewart, W. E., and Lightfoot, E. N. *Transport Phenomena.* Wiley, 1960.

Crank, J. *Mathematics of Diffusion.* Oxford, 1956.

Danckwerts, P. V. *Gas-Liquid Reactions.* McGraw-Hill, 1970.

Pigford, R. L., and Sherwood, T. K. *Absorption and Extraction.* McGraw-Hill, 1952.

Treybal, R. *Mass Transfer Operations.* McGraw-Hill, 1968.

chapter 8

Mass Transfer in Living Systems

INTRODUCTION

The general principles presented in Chapter 7 will be applied here to several situations that involve mass transfer in living systems. Living organisms, which must be supplied with nutrients and must dispose of waste products to sustain life, are naturally systems in which mass transfer and diffusion are of critical importance. An understanding of these processes is certainly desirable if an understanding of the factors governing the operation of living systems is to be gained.

In many of the control mechanisms of the human body, a primary mechanism involves variation of the permeability of membranes to selected solutes. This adjustment of the diffusion properties of materials, stimulated by chemical agents, is best understood by an examination of the fundamental flux expressions which apply.

In the design and operation of artificial organs, which are supplementing or replacing bodily functions, mass transfer quite often is the major problem and the key process that must be controlled. The application of the principles of the previous chapter are of special importance in this area.

In addition to concentration-driven diffusion, there are several other chemical and physical processes that either result in or result from gradients in concentration, free energy, and hydrostatic pressure. Some of the more important of these are summarized below.

Osmosis. This process, discussed in Chapter 1, differs primarily from ordinary diffusion in that only the solvent, rather than the solute, is able to penetrate the separating membrane. Usually, the solute molecules that cause the osmotic driving force are unable to pass through the membrane because of their size. In ordinary diffusion, all species are able to penetrate the membrane, although the permeability may be different for each species.

Filtration. This process, by which a suspension can be separated into a filtrate phase relatively free of larger particles and a filtrand phase relatively more concentrated in larger particles, is a mechanical process driven by a pressure difference across a filter medium such as a membrane. In living systems, filtration is often accompanied by osmosis.

Active transport. A not completely understood process occurs in several critical locations in living systems. This process results in a species being transferred across a membrane against a concentration gradient, that is, "up the concentration hill." We realize in light of free-energy considerations that such a process must be supplied energy from an external source in order to operate. The concentration of stomach acid as discussed in Chapter 1 is an example of a situation where active transport takes place.

Pinocytosis. The mechanical process that results in large molecules being engulfed in the cell membrane and subsequently transported to the interior is a primary process in supporting cell function. The driving forces for this process are not completely understood.

Dialysis. This term is included here, not because it refers to a different mode of transfer or to an exclusively physiological process, but because it is important in life support processes and artificial organs. It refers to the concentration-driven diffusion of a solute across a permeable membrane, as demonstrated in Example 7.9.

Facilitated diffusion. When the diffusion of a chemical species is accompanied by chemical reaction involving that species, as in Examples 7.4 and 7.5, the net rate of transport is often increased or "facilitated." The diffusion of oxygen in hemoglobin solution is an example of this mechanism of mass transfer.

These various mechanisms will be discussed further in this chapter in the context of specific examples. The application of basic material balance relations in conjunction with the treatment of mass transfer rates will be presented for a selected group of physiological subsystems.

Example 8.1. Mass Transfer in the Respiratory System

The human respiratory system is a classic example of a mass transfer process involving convection, diffusion, and chemical reaction in a two-phase gas-liquid configuration. Although respiratory physiologists have historically generated a voluminous amount of research information about the system, the mass transfer aspects have only recently been fully appreciated.

We recall Example 2.1 that developed the material balance relations for a normal adult lung. Some of the important results of the material balance are summarized in Table 8.1.

Table 8.1. Lung Material Balance Information

Overall ventilation rate (BTP)	6,000 ml/min	
Alveolar ventilation rate (BTP)	4,200 ml/min	
Pulmonary blood flow rate	5,000 ml/min	

	CO_2	O_2
Metabolic rates (STP)	200 ml/min	250 ml/min
Arterial blood concentration (STP)	0.48 ml/ml	0.195 ml/ml
Venous blood concentration	0.52 ml/ml	0.145 ml/ml
Alveolar gas concentration	0.050 ml/ml dry gas	0.153 ml/ml dry gas
Partial pressure in alveoli	40 mm Hg	100 mm Hg
Partial pressure–venous blood	46 mm	40 mm
Partial pressure–arterial blood	40 mm	100 mm

Figure 8.1 shows some typical equilibrium curves for oxygen and carbon dioxide concentrations versus partial pressure in whole human blood. This figure also indicates the interrelation between CO_2 and O_2 in the equilibrium relationships described by physiologists as the Haldane and Bohr effects. It should also be pointed out that these plots of concentration versus partial pressure are presented inversely from plots generally found in engineering and physical chemistry texts; that is, the ordinate and abscissa are reversed. This is an established practice in physiology and stems from the universal use of partial pressure as the independent variable.

It is important here to note that the O_2 and CO_2 are held in chemical combination in the red cell as well as in physical solution in the red cell and in the plasma. Table 8.2 summarizes the important overall chemical reactions and modes of associations in the system.

While O_2 is carried almost exclusively by chemical combination, CO_2 is also carried extensively by bicarbonate ions, both in the red

Table 8.2. Gas-Carrying Mechanisms

Oxygen:	
Red cell	
$O_2 + H^+Hb \rightleftharpoons HbO_2 + H^+$	~99% as HbO_2
$O_2 + HbCO_2 \rightleftharpoons HbO_2 + CO_2$	~1% in solution
Carbon dioxide:	
Red cell	
$CO_2 + HbO_2 \rightleftharpoons HbCO_2 + O_2$	~20%
$CO_2 + H_2O \rightleftharpoons H_2CO_3 \rightleftharpoons HCO_3^- + H^+$	~70%
Plasma	
$CO_2 + H_2O \rightleftharpoons H_2CO_3 \rightleftharpoons HCO_3^- + H^+$	~10%

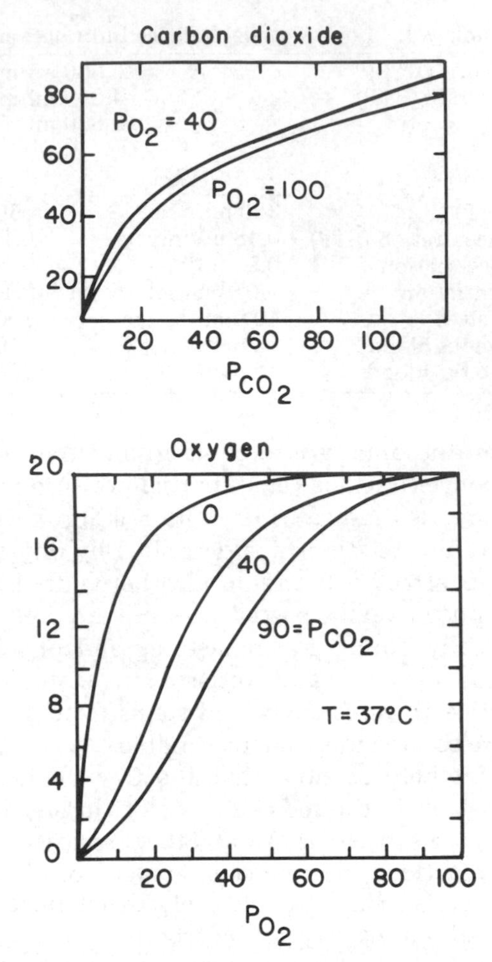

Fig. 8.1. Solubility data for oxygen and carbon dioxide. Ordinate values are ml dissolved gas (STP)/100 ml blood.

cell and in plasma. This helps to explain why the carrying capacity for CO_2 for normal arterial blood is over twice that for O_2. Although almost all the hemoglobin is used for carrying O_2, there is considerable capacity remaining for CO_2. These reactions in Table 8.2 also give some idea of why there is coupling or interdependence between the equilibrium curves in Figure 8.1.

Another word is also necessary about the ordinates in Figure 8.1, which are given as volume percent O_2 and CO_2, that is, the equivalent volume in ml of CO_2 and O_2 that is physically or chemically carried per ml of blood. Both these quantities presume a certain concentration of hemoglobin in the red cells and a certain fraction of red cells in the blood. Table 8.3 gives some sample typical values of the physical and chemical properties of human blood which are consistent with Figure 8.1.

Table 8.3. Some Properties of Blood

Property	Value
Hemoglobin content, gm/100 ml RBC	34.0
Hemoglobin content, gm/100 ml blood	15.6
Hematocrit (volume fraction RBC)	0.46
Density	1.06 gm/ml
Surface tension	47 dynes/cm
Viscosity:	plasma—1.4 cp
	H = 0.40—2.4 cp

Since the molecular weight of the protein hemoglobin is about 68,800 and each molecule of hemoglobin by virtue of its associated four exposed iron atoms can carry four molecules of O_2 at saturation, it can easily be computed that 1 gm of pure hemoglobin can carry the equivalent of 1.32 ml of O_2 (STP) at 100% saturation. The maximum saturation value, using the data of Table 8.3, is then

$$\frac{1.32 \text{ ml } O_2}{\text{gm Hb}} \times \frac{15.6 \text{ gm Hb}}{100 \text{ ml blood}} = \frac{20.5 \text{ ml } O_2}{100 \text{ ml blood}}$$

Added to this is the amount that can be carried in physical solution as dissolved in the plasma. This will be a function of the total pressure of O_2 present (p_{O_2}). For the data of Table 8.1 this amounts to 0.5 ml/100 ml blood.

A more fundamental way to present the relationships shown in Figure 8.1 would be to use percent saturation as the ordinate, that is, the fraction O_2 capacity at that partial pressure/O_2 capacity at saturation. Many texts use this presentation. Once the hemoglobin content is known, however, the actual concentration ordinate as used in Figure 8.1 is much more convenient for material balance and mass transfer calculations.

We are primarily concerned in this chapter with the process of mass transfer. For the lung, we are interested in characterizing the process whereby O_2 is delivered from the gas phase to incoming venous blood to increase its O_2 content, simultaneously removing CO_2 from venous blood to lower the content of that substance. Figure 8.2 schematically shows the major components that separate alveolar gas from the red cell, along with a qualitative O_2 partial pressure indication and indicates that the O_2 must be transported

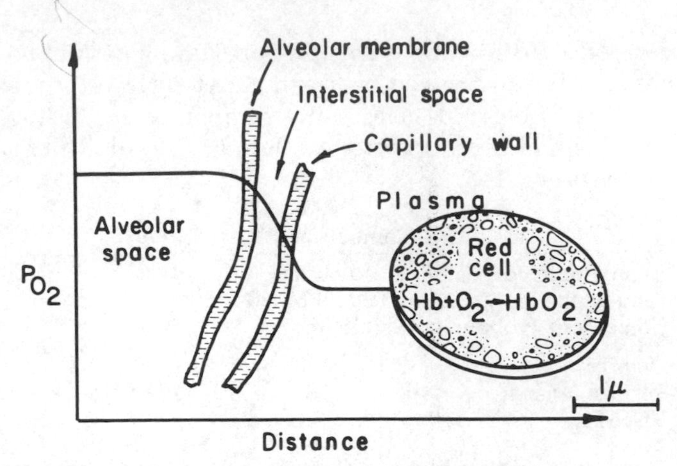

Fig. 8.2. Alveolar-capillary morphology.

through a series of resistances, summarized as follows:

1. The alveolar gas film.
2. The alveolar membrane.
3. The interstitial space.
4. The pulmonary capillary wall.
5. The film of plasma near the wall.
6. The plasma.
7. The red cell membrane.
8. The intercellular fluid.
9. Chemical reaction.

This is a total of nine distinct steps. However there is good evidence that steps 1, 7, and 8 present negligible resistance. Steps 2, 3, and 4 can be considered as one resistance. Steps 5 and 6 can also be combined, so that the simplified process now is:

1. Diffusion through the membrane.
2. Diffusion through the plasma.
3. Reaction in the cell.

The molar flux of O_2 diffusing through any layer may now be written (letting $A = O_2$) as

$$J_A = -C\, \mathfrak{D}_{AM}\, dY_A/dz \quad \text{moles } O_2/\text{area time} \tag{8.1}$$

Since we usually deal with the volumetric equivalents rather than with moles, we multiply through by the appropriate molar volume $\widetilde{V}$ and write

$$\widetilde{V} J_A = -C\, \widetilde{V}\, \mathfrak{D}_{AM}\, dY_A/dz \quad \text{vol } O_2/\text{area time} \tag{8.2}$$

Noting that $C\,\widetilde{V} = 1$, and since we want the rate rather than the flux, we multiply through by S (the surface area) and get

$$\dot{V}_{O_2} = S\,\widetilde{V}\,J_A = -S\,\mathfrak{D}_{AM}\,dy_A/dz \tag{8.3}$$

In the gas phase,

$$y_A = p_{O_2}/P$$

so

$$\dot{V}_{O_2} = \frac{-S\,\mathfrak{D}_{AG}}{P}\,dp_{O_2}/dz \tag{8.4}$$

Across a thin membrane the derivative can be approximated, so

$$\dot{V}_{O_2} = \frac{-S\,\mathfrak{D}_{AM}\,\Delta p_{O_2}}{P\,\Delta z} \tag{8.5}$$

Within a liquid phase this is written as

$$\dot{V}_{O_2} = \frac{-S\,\widetilde{V}\,\mathfrak{D}_{AL}}{H^*\,\Delta z}\,\Delta p_{O_2} \tag{8.6}$$

where H^* is the local prevailing equivalent of Henry's law H.

When the transfer takes place across a liquid-gas interface separated by a membrane, as in Figure 8.2, (8.5) and (8.6) may be combined to write

$$\dot{V}_{O_2} = \left(\frac{1}{\dfrac{P\,\Delta z_g}{\mathfrak{D}_{Ag}\,S} + \dfrac{\phi\,z_m}{\mathfrak{D}_{AM}\,S} + \dfrac{C\,H^*\,\Delta z_l}{\mathfrak{D}_{AL}\,S}} \right) \Delta p_{O_2} \tag{8.7}$$

This equation could be used to describe the rate of uptake by blood of oxygen at any point in the lung, if the following assumptions were made.

1. The chemical reactions involved are at least an order of magnitude faster than the diffusion steps.
2. The lung is homogeneous throughout as far as its mass transfer aspects are concerned.
3. The total area for transfer S is known.

Then the symbols in (8.7) would be interpreted as follows:

P = total pressure
Δz_g = gas film thickness
$\mathfrak{D}_{Ag}$ = diffusivity of oxygen in the gas phase
S = total area for mass transfer
ϕ = membrane distribution coefficient
z_m = total membrane thickness
$\mathfrak{D}_{AM}$ = diffusivity of oxygen in the membrane
C = total molar concentration of blood

H^* = slope of dissociation curve (Figure 8.1)
Δz_l = plasma film thickness
$\mathfrak{D}_{Al}$ = diffusivity of oxygen in plasma

Physiologists have considerably simplified this expression with the use of the diffusing capacity D_{O_2}, defined by

$$\dot{V}_{O_2} = D_{O_2} \, \Delta p_{O_2} \tag{8.8}$$

which, although having the virtue of simplicity, tends to obscure the physical processes and parameters that are actually involved. Equation 8.7, which is already greatly simplified, shows that the flow-pressure relationship contains eleven distinct parameters, all of which influence the transfer rate. In other words, the diffusing capacity is a function of the eleven variables listed above. An understanding of (8.7) will permit comparison of diffusing capacities of different species on a more valid physical ground than is possible from standard physiological texts.

As indicated repeatedly in Chapter 7, many quantities in (8.7) are difficult to measure directly. The use of mass transfer coefficients are again appropriate. We may define the following:

$$k_g = \mathfrak{D}_{Ag}/\Delta z_g \qquad k_g a = \mathfrak{D}_{AG} \, S/\Delta z_g V \tag{8.9a}$$

$$k_m = \mathfrak{D}_{Am}/z_m \qquad k_m a = \mathfrak{D}_{Am} \, S/z_m V \tag{8.9b}$$

$$k_l = \mathfrak{D}_{Al}/\Delta z_l \qquad k_l a = \mathfrak{D}_{Al} \, S/\Delta z_l V \tag{8.9c}$$

So that (8.7) becomes

$$\dot{V}_{O_2} = ka \; V \, \Delta p_{O_2} \tag{8.10}$$

where

$$ka = 1/\Sigma_i \; \psi_i/ka_i \tag{8.11}$$

where

$$\psi_i = P, \phi, \text{ or } C \, H^*$$

and

$$ka_i = k_g a, \; k_m a, \; k_l a \tag{8.12}$$

Example 8.2. Oxygen Uptake in the Pulmonary Capillary

Although the alveolar gas is well mixed, at least locally, the blood in the pulmonary capillaries becomes progressively richer in oxygen content as it passes along the length of the capillary (as shown schematically in Figure 8.3), so that normal conditions dictate that the equivalent partial pressure of O_2 changes from 40 mm Hg at the beginning of its exposure to 100 mm Hg at the end. Equation 8.10 can now be used in its differential form to give the uptake rate at any

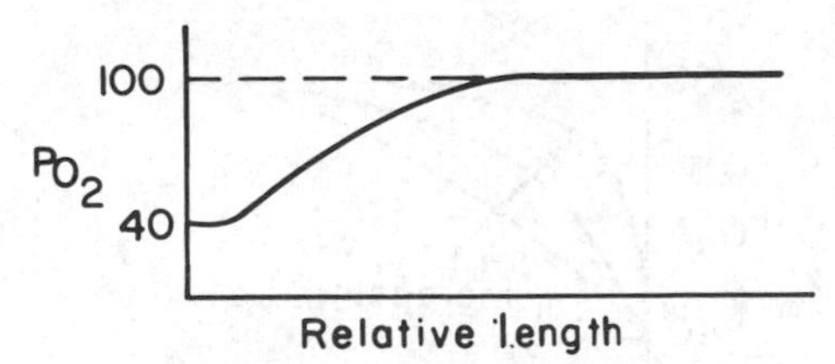

Fig. 8.3. Relative longitudinal capillary profile.

point along the capillary as

$$d\dot{V}_{O_2} = ka\, dV\, \Delta p_{O_2} \tag{8.13}$$

where the Δ now refers to the alveolar p_{O_2} – capillary p_{O_2} difference. This difference goes from 60 mm near the beginning to near zero at the exit. The differential volume dV can be approximated by Adz, where A is the area for mass transfer per unit length of capillary. Then the material balance equation that holds in the capillaries may be written, by analogy with (5.22) in Example 5.5 or by applying a material balance to a differential length dx of capillary,

$$Q_B\, dx_{O_2}/dz = ka\, A\, \Delta p_{O_2} \tag{8.14}$$

where

Q_B = volumetric flow rate of blood in the capillary
x_{O_2} = O_2 content of blood, ml O_2/ml blood

This relation simply says that the rate at which the capillary blood picks up O_2 as it flows along is equal to the rate at which O_2 is supplied by mass transfer into the capillary.

Since $x_{O_2} = f(p_{O_2})$ from Figure 8.1, it is now possible in principle to integrate (8.14) with the initial venous condition as a boundary condition and to produce the exponential curve shown in Figure 8.3 as a solution. This may indeed be done, but it must be done numerically, due to the sigmoid shape of the equilibrium relation $x = f(p)$ over the range of partial pressures from 40 to 100 mm Hg. Some typical results are shown qualitatively in Figure 8.4, which also shows the effects of flow rate and overall transfer coefficient on the resulting longitudinal concentration profiles.

As can be seen from Figure 8.4, any factors which will reduce the overall ka, which would include such factors as reduction in area for transfer, increase in membrane thickness, or changes in solubility, would cause a lowering of the O_2 uptake rate. Also, when the flow rate through the capillary is increased, unless there is a corresponding change in the overall ka, the O_2 uptake will be lessened. This would

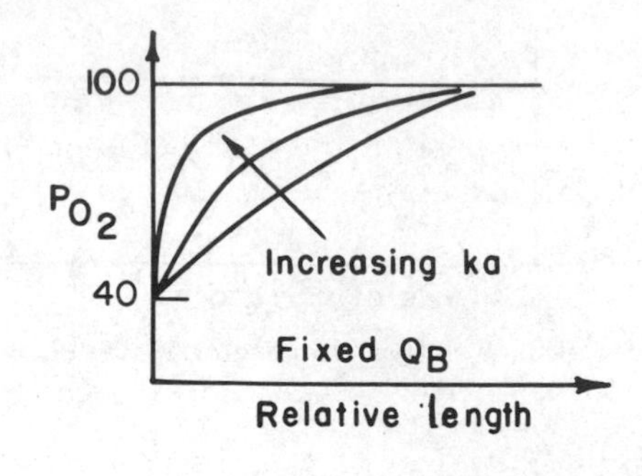

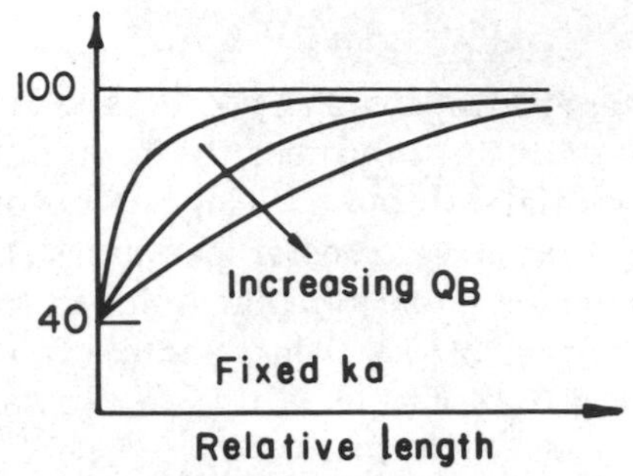

Fig. 8.4. Mass transfer aspects of the pulmonary capillary.

partially be compensated by the increased flow. In situations such as exercise, the overall *ka* increases along with the blood flow rate, although the mechanisms are not completely understood. Most likely, the increase in *ka* is due primarily to an increase in area and by some lowering of the liquid side-film resistance, as would be predicted from the correlations discussed in Chapters 5 and 7.

The effect of the chemical reaction rate in the red cell has been neglected up to this point. One way to take this rate into account in the overall rate of transfer is to consider the chemical reaction as an added resistance and to include it conceptually in k_l as defined in (8.9c). Another way would be to actually include a reaction term in the overall material balance equation.

Perhaps the best way to take the chemical reaction into account in an overall fashion is to consider it as an additional resistance, as suggested by Roughton. The overall mass transfer coefficient corrected for the effect of the chemical reaction can then be expressed as

$$1/ka = 1/ka' + 1/\theta\, V_c \tag{8.15}$$

where

ka' = overall mass transfer coefficient without the reaction

V_c = volume of blood in the lung capillaries
θ = rate of uptake of gas by red cells, ml O_2/ml blood mm Hg (based on a 1 mm partial pressure difference between red cell interior and plasma)

In a normal adult lung, V_c is about 75 ml and Q_B is about 100 ml/sec. The residence time of the blood in the pulmonary capillaries is then about 0.75 sec. The overall kaV (or D) is then about 20 ml/min mm Hg for O_2 for resting adults, increasing to about 65 ml/min mm during exercise. For CO_2, the value of H^* in (8.7) is much lower, and the resting value of D is about 400 ml/min mm, increasing to about 1,200 during exercise. Although CO_2 is referred to as "diffusing" 20 times faster under these conditions, it should be pointed out that this is due to the difference in solubility and the slope of the partial pressure concentration equilibrium curve as shown in Figure 8.1, and not due to any gross difference in the values of the diffusion coefficients $\mathfrak{D}_{im}$. Investigation of (8.6) will explain this factor of 20 more conclusively and more concisely. The local value for H^* for the two gases under conditions in the lung may be estimated from Figure 8.1. H^* is defined by

$$p_1 = H_i^* x_i \tag{8.16}$$

where x_i = ml of O_2 or CO_2/ml of blood and may be estimated as

$$H_i^* = dp_i/dx_i \tag{8.17}$$

for each component from Figure 8.1 by taking the slopes of the solubility or equilibrium curves.

For O_2:

$$H^* = dp_i/dx_i = 60/0.04 = 1200$$

For CO_2:

$$H^* = dp_i/dx_i = 6/0.04 = 150$$

This rough estimate of the average value of H^* over the normal range of operation shows an order of magnitude difference in the solubility or capacity characteristics and gives insight into the major physical differences causing the higher values of diffusing capacities for CO_2.

The material presented in this section is especially valuable when one sets out to develop a mathematical model for the respiratory system. Figure 8.5 shows the simplest two-compartment model used to describe transient changes in the amounts of O_2 and CO_2 present in the respiratory system. One of the most important limitations placed on these types of early two-compartment models was that the blood leaving the lungs (that, is arterial blood) was in equilibrium

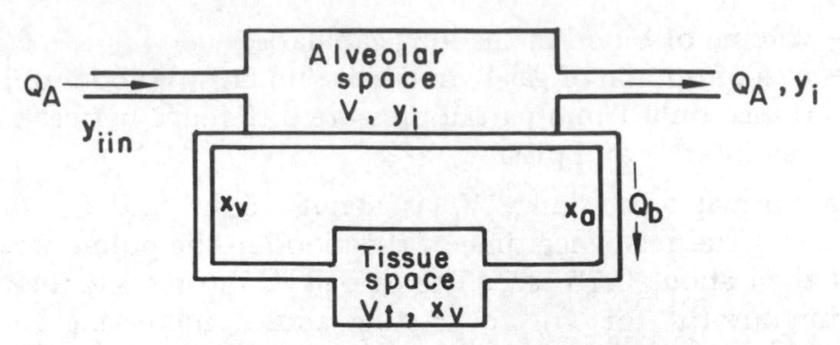

Fig. 8.5. Simple respiratory model.

with the gas in the lungs. In another sense, the lung in this model is a perfect mass transfer device. Then the model equations become, from (2.7) and (2.8),

Alveolar space:

$$dy_i/dt = Q_A(y_{iin} - y_i)/V + Q_b(x_v - x_a)/V \tag{8.18}$$

Tissue space:

$$V_t\, dx_v/dt = Q_b(x_a - x_v) + R\,M\ (-M \text{ for } O_2) \tag{8.19a}$$

Equilibrium relation:

$$x_a = x_a(y_i) \tag{8.19b}$$

where

y_i = mole fraction in gas phase
y_{iin} = mole fraction in inspired gas
x_v = volume fraction in venous blood, ml gas/ml blood
x_a = volume fraction in arterial blood, ml gas/ml blood
Q_A = alveolar ventilation rate, ml/min
Q_b = blood flow rate, ml/min
V = lung volume, ml
V_t = equivalent tissue space volume, ml
R = respiratory quotient
M = metabolic generation rate, ml O_2 consumed/min

These model equations, when solved simultaneously, can quite nicely give both transient and steady-state material balance information for normal lungs. In early respiratory modeling these types of equations were used along with a feedback control loop to control the ventilation rate, and phenomena such as the response to CO_2 inhalation were modeled. If the mass transfer in the lung is impeded, however, so that the exiting arterial blood is not in equilibrium with the gas in the lung, then this model is insufficient and must be modified. The addition of another pool to follow pulmonary blood, and a fourth pool to time-average pulmonary blood and remix shunted blood, will allow the incorporation of quantitative description of the mass transfer processes which have been discussed in this section. Figure 8.6 shows the modified mass transfer model.

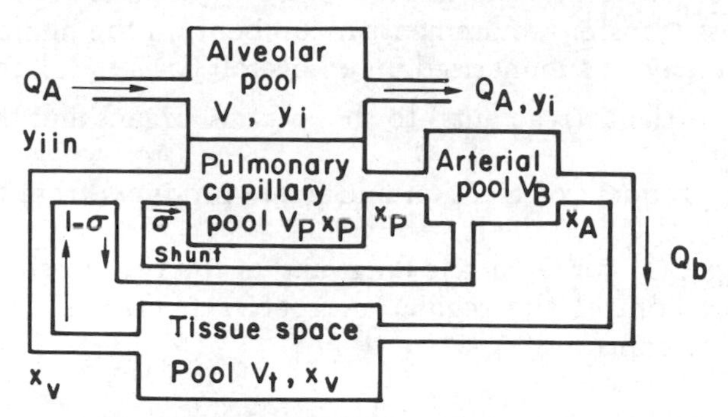

Fig. 8.6. More detailed respiratory model.

Material balance equations are now written as before for the additional two pools, and the mass transfer from the alveolar pool to the pulmonary pool is described by the use of an overall mass transfer coefficient. The following model equations can now be used to describe the effects of impeded mass transfer due to increased resistance, decreased area, or excessive shunting.

Alveolar pool:

$$dy_i/dt = Q_A\,(y_{iin} - y_i)/V - ka(y_i - y^*) \tag{8.20}$$

Pulmonary pool:

$$V_p\ dx_p/dt = Q_b(\sigma x_v - x_p) + ka(y_i - y^*)\ V \tag{8.21}$$

Arterial pool:

$$V_b\ dx_A/dt = Q_b((1 - \sigma)\,x_v + \sigma\,x_p - x_A) \tag{8.22}$$

Tissue pool:

$$V_t\ dx_v/dt = Q_b\,(x_a - x_v) + R\,M \tag{8.23}$$

Equilibrium relation:

$$y^* = y^*(x_p, x_v) \tag{8.24}$$

The reader will note that at normal steady-state conditions (8.20) and (8.21) when summed will produce (8.18). The symbols in the above equations are defined for the earlier model and in Figure 8.6. The fraction not shunted is represented by σ, and the representative driving force concentration difference by $y_i - y^*$, where y^* is in equilibrium with the average of x_v and x_p.

These model equations must be solved simultaneously. One popular method is to employ an analog computer. The interested reader should consult the references listed in the bibliography for further results and details concerning these modeling techniques.

Example 8.3. Mass Transfer in the Circulatory System

The mass transfer requirements incumbent on the human circulatory system may be summarized on an overall basis as follows:

1. Supply nutrients (reactants) to the internal organs and the cells of the body.
2. Remove products of reaction and waste products from these cells and organs.
3. Exchange CO_2 for O_2 in the lungs and at the reactor sites.
4. Transport control and regulatory species from their production sites to the control sites.

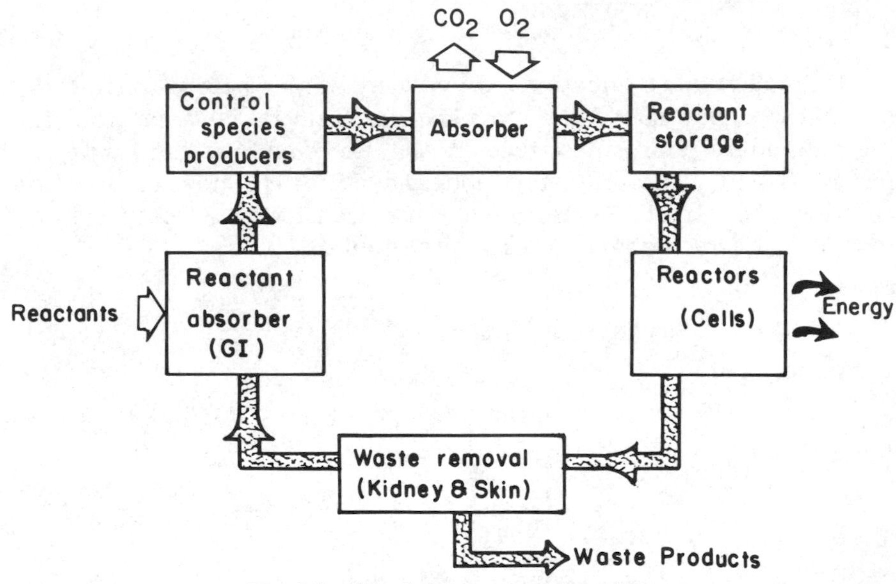

Fig. 8.7. Circulatory system functions.

Figure 8.7 summarizes these mass transfer tasks. From a chemical engineering viewpoint the control and regulation of this system, as schematically depicted in Figure 8.7, is the most interesting, since control is mostly effected by the regulation of the mass transfer characteristics of the system, either by the variation of the permeabilities (or diffusivities) of membranes, by the regulation of species concentrations, by changing the net rate at which chemical reactions occur, or by combinations of these. It is useful therefore to develop some quantitative relations for the amount of various species present in the system so that the relative importance of these phenomena and of their effects on the overall system can be elucidated.

In general, we are concerned about the transfer of any given chemical species between pairs of blocks in Figure 8.7. We may begin the analysis by writing the material balance for any given compartment in Figure 8.7, again from (2.7) or (2.8),

$$V_j \, dC_{ij}/dt = Q_{Bj} \, (C_{iB\,\mathrm{in}} - C_{iB\,\mathrm{out}}) + \dot{R}_{ij} \pm Q_{Dj} \, C_{iD} \tag{8.25}$$

and the associated mass transfer equation:

$$Q_{Bj} \, (C_{iB\,\mathrm{in}} - C_{iB\,\mathrm{out}}) = N_{ij} \, S_{ij} \tag{8.26}$$

where

Q_{Bj} = blood flow rate to compartment j, ml/min
C_{ij} = concentration of species i in compartment j, meq/ml
N_{ij} = mass transfer flux in compartment j, meq i/area min
S_{ij} = mass transfer surface area for i in j, area
$\dot{R}_{ij}$ = net production rate of i in j, meq/time
V_j = volume of compartment j
Q_{Dj} = flow rate of other flows out of the compartment j, ml/min
C_{iD} = concentration of i in other flows

A rate equation may also be written:

$$N_{ij} = K_{ij} \, \Delta C_{ij} \tag{8.27}$$

where

K_{ij} = the overall mass transfer coefficient for i in j

Equation 8.25 states that the accumulation of species i in compartment j is equal to the net input of i minus the net outflow of i plus the rate at which i is formed within the compartment. Equation 8.26 equates the net rate of removal of i by mass transfer across the compartment boundaries with the net rate of mass transfer from the compartment to the bloodstream. Equation 8.27 equates the mass transfer rate to the product of the overall mass transfer coefficient and the overall concentration driving force from compartment to bloodstream. The reader should note the resemblance of this general balance formulation to the model equations for the respiratory system delineated earlier.

Example 8.4. Urea Transport in the Circulatory System

Urea, a product of protein metabolism, is produced in cells, carried through the circulatory system, extracted in the kidneys, and disposed of through the urinary system. We can describe this process by writing overall material balance equations for each of the major compartments concerned. We will drop the subscript i and understand that C denotes urea concentration.

Metabolic pool (cells):

$$V_m \, dC_m/dt = Q_{Bm} \, (C_{B\,\mathrm{in}} - C_{B\,\mathrm{out}}) + \dot{R}_m \tag{8.28}$$

Kidneys:

$$V_k \, dC_k/dt = Q_{Bk} \, (C_{B \text{ in}} - C_{B \text{ out}}) - Q_u C_u \tag{8.29}$$

Bladder:

$$\frac{d}{dt} (V_u \; C_u) = Q_u \; C_u - Q_v \; C_u \tag{8.30}$$

The blood flow term through the metabolic pool can be more appropriately represented by use of (8.26) as

$$Q_{Bm} \, (C_{B \text{ in}} - C_{B \text{ out}}) = N_{mB} \; S_{mB} = K_{mB} \; S_{mB} \, \Delta C_{Bm} \tag{8.31}$$

At steady state, that is, no transients in the cells or kidneys,

$$Q_{Bm} \, (C_{B \text{ in}} - C_{B \text{ out}}) + \dot{R}_m = 0 \tag{8.32}$$

$$Q_{Bk} \, (C_{B \text{ in}} - C_{B \text{ out}}) - Q_u \; C_u = 0 \tag{8.33}$$

$$dV_u/dt = Q_u - Q_v \tag{8.34}$$

When connecting overall material balance equations are written for more than one pool, one must take care to differentiate the in and out concentrations for each pool. In this case $C_{B \text{ out}}$ for the cells is the $C_{B \text{ in}}$ for the kidneys and vice versa.

Since $\dot{R}_m$ is of the order of magnitude of 1 gm urea/hour, or 16 mg/min, the change in concentration $C_{B \text{ in}} - C_{B \text{ out}}$ through these compartments would seem to be practically negligible, being on the order of 0.30 mg%. However, under normal conditions, the plasma concentration of urea is only on the order of 2.5 mg%, so the change through the compartment represents about 10% of the operating concentration level of urea.

In the kidney pool, however, Q_{Bk} is on the order of 1,300 ml/min, so the urea concentration change is on the order of 1 mg% through the kidney compartment. Since the concentration of urea in urine is about 16 mg/ml, (8.33) tells us that the flow rate of urine is on the order of 1 ml/min. Equation (8.34) then says what the relation between bladder volume and voiding flow Q_v is under normal conditions.

The mass transfer relations which accompany the material balance equations, such as (8.31), are of considerable interest. Equation (8.31) tells the rate at which urea can be absorbed by the bloodstream from the metabolic or cellular pool. K_{mB} is the overall mass transfer coefficient for transfer from the metabolic pool to the bloodstream, while S_{mB} represents the available effective interfacial area for mass transfer between these two pools. As in previous situations, the effective area is very difficult to determine, and so the usual procedure once again is to define a new coefficient as

$$K_{mB} \; S_{mB} = Ka_{mB} \; V_m \tag{8.35}$$

where V_m is the volume of the metabolic or cellular pool. The symbol a_{mB} represents the specific area, that is, the area per unit volume. The combination Ka_{mB} is now used and measured directly.

The mass transfer processes that take place inside the kidney itself will be discussed in a later section.

The overall mass transfer coefficient Ka_{ij} in the circulatory system must in general describe the joint contributions of three separate mass transfer processes that carry material across the capillary membranes in the microcirculation and ultimately across the cell membranes in the tissues in general. These processes may be briefly summarized as follows:

1. *Concentration-driven diffusion* across the capillary wall, through the interstitial fluid, and across the cell membrane.
2. *Convective flow* of solute, along with the bulk fluid which flows hydrodynamically as the result of a net pressure differential.
3. *Osmosis*, resulting from the osmotic pressure gradient, caused by a solute concentration difference in turn caused by a difference in large molecule concentration difference across the capillary wall—usually protein molecules which are on the average too large to permeate the capillary walls at a rate comparable to electrolytes and water. This back osmotic pressure tends to militate against the hydrostatic pressure differential and affects the convective flow mentioned above.

Perhaps the easiest way to quantify these processes is to combine 2 and 3 and refer to the net convective flow resulting from the net overall pressure differential, that is, the sum of the hydrostatic and osmotic pressure differentials. Then the total flow of species i across the capillary wall can be expressed, using the standard flux relationship from Chapter 7, as

$$N_i = x_i \sum_i N_i - C\,\mathfrak{D}_{im}\,dx_i/dz \qquad (8.36)$$

with

$$\sum_i N_i = C\,v = C\,K\,\Delta p \qquad (8.37)$$

where

K = flow permeability of the capillary membrane, moles/area time mm Hg

Δp = net pressure difference across the capillary wall

The diffusion flux $C\,\mathfrak{D}_{im}\,dx_i/dz$ across the capillary wall can be written as $-\mathfrak{D}_{im}\,\Delta C_i/\Delta z$, so that the overall flux expression becomes

$$N_i = C_i\,K\,\Delta p + \mathfrak{D}_{im}\;\Delta C_i/\Delta z \qquad (8.38)$$

so that it is evident that the flux of a species across the capillary wall is the sum of a flux driven by pressure and a flux driven by concentration.

The flux through the interstitial fluid and finally into the intercellular fluid at steady state are both equal to the flux written above. Figure 8.8 sums up the flux relationships.

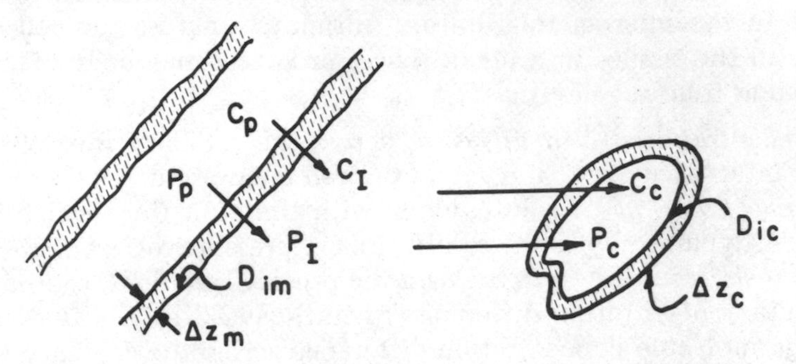

Fig. 8.8. Capillary tissue mass transfer. $N_i = C_p K_1 (p_p - p_I) + \mathfrak{D}_{im}(C_p - C_I)/\Delta z_m = C_I K_2 (p_I - P_c) + \mathfrak{D}_{ic}(C_I - C_c)/\Delta z_c$.

It is important to note two significant points about the fluxes in Figure 8.8. First, they can take place in either direction; in fact there is evidence to suggest that the total convective flux varies in direction along the length of a single section of capillary in the microcirculation, being directed outward near the beginning and inward at the end (or venous end). Products of cellular reactions, such as urea, which are being removed by the circulatory system, must be undergoing a net transfer into the capillary, mostly by diffusion, while reactants or fuel for the cells must be undergoing a net transfer outward. Also, the permeability of the membranes involved, both the flow permeability K and the chemical permeability $\mathfrak{D}_i/\Delta z$, can evidently be controlled by the secretion of regulatory agents such as hormones and enzymes carried in the bloodstream. In addition (perhaps most importantly) variations in the concentration of protein constituents in the plasma in turn change the net osmotic pressure difference and consequently the net pressure differences across the capillary and cell walls. This naturally has a profound effect on the magnitude and direction of the convective flow at any given point in the system.

No discussion of the mass transfer phenomena relating plasma and interstitial fluid in the microcirculation can be complete without consideration of *Gibbs-Donnan equilibrium*. The Gibbs-Donnan rule may be directly deduced from free-energy considerations as discussed in Chapter 1. It states that under equilibrium conditions the product of the concentrations of any pair of diffusible cations (for example, Cl^-, $PO_4^{\equiv}$) and anions (Na^+, K^+) on one side of a membrane must equal the product of concentrations of the same pair on the other side. When nondiffusible ions (such as larger molecules) are present, the diffusible ions must therefore be distributed unequally at equilibrium to satisfy this rule. The other constraint to be obeyed is that on any side of the membrane the number of cations and anions must be equal. The combination of these two effects or constraints will then produce an unequal number of total ions across the membrane and therefore cause the net osmotic pressure difference referred to above.

The general area of electrolyte and ionic balance and the associated problems of acid-base regulation are subjects of great interest and complexity. While equilibrium considerations must be evaluated carefully when a nonequilibrium situation such as mass transfer is occurring, the interested student is urged to consult either Best and Taylor's classic text on this subject or Pitts's *Physiology of the Kidney and Body Fluids* for further treatment of these topics.

Example 8.5. Mass Transfer in the Kidney

The human kidney is perhaps the most interesting example of a combination of the various modes of mass transfer that can be found in nature. The processes of osmosis, filtration, convection, diffusion, and active transport are all important mass transfer processes that apparently occur in the kidney and in fact constitute almost its entire function. We shall examine each of these processes in a semiquantitative fashion as they occur in the kidney, in the hope of strengthening not only the reader's understanding of mass transfer but also his understanding of the elements of normal renal function and its associated control and regulation.

Figure 8.9 schematically represents the major mass transfer processes as they occur in the nephron, according to the currently accepted hypotheses. About one million of each of these elements are present in each human kidney, and the blood flow through the renal artery leading to each kidney is distributed throughout, so that each individual nephron receives only a tiny fraction of the total flow. About 500 ml/min of blood normally perfuses each kidney. The flow rates used in the following analysis of the mass transfer aspects of the individual nephrons will be in terms of the overall flow rates to each kidney.

The major physiological functions of the kidney as a mass transfer device are the regulation of the composition of the bloodstream

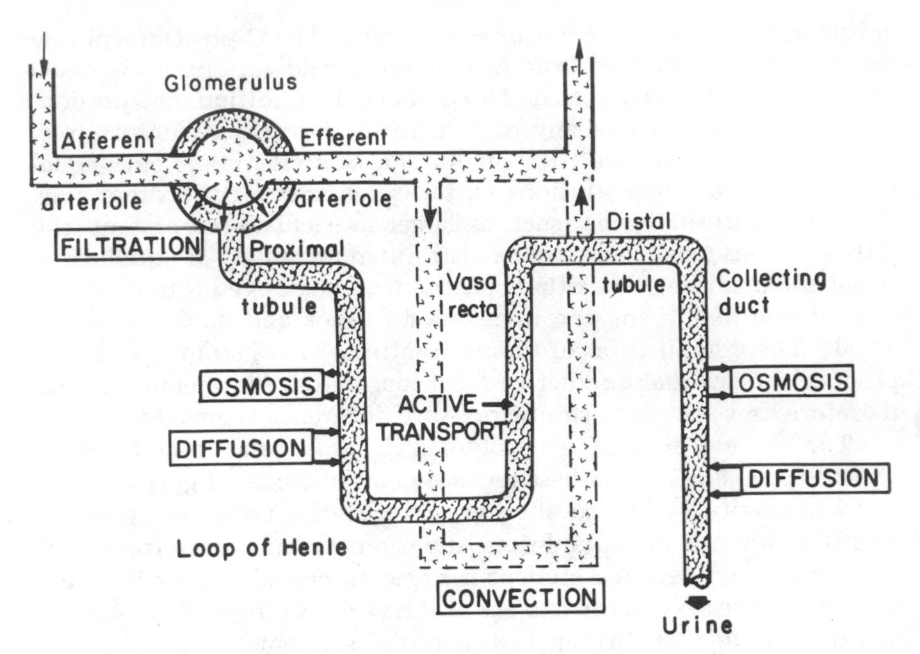

Fig. 8.9. Mass transfer in the nephron.

and hence the body fluids and regulation of the volume of the extracellular fluids of the body. While the kidneys are not the sole mechanism for accomplishing this, they do play a critical role in the removal of certain products of the metabolic energy conversion process, such as protein metabolites and acidic radicals like phosphate and sulfate. By varying the total amount of urine produced, they also act as fluid-volume control elements. Their entire function in these two tasks is acting as mass transfer devices.

The first separation process occurs in the glomerulus where blood plasma is filtered through the glomerular membrane at an overall rate of about 150 ml/min. The driving force for this filtration process is a net pressure gradient across the glomerular membrane, which contains a very large number of pores averaging in humans about 50 angstroms in diameter and 500 angstroms in length. The relatively small diameter of these pores prevents red cells (about 5 microns in diameter) and very large protein molecules from being filtered. As a result there is a back osmotic pressure of about 25 mm Hg caused by the concentration difference between filtrate and filtrand. The hydrostatic pressure difference across the glomerular membrane is normally about 50 mm Hg, so that the net pressure difference to force plasma through the membrane is about 25 mm Hg. It has been observed that this pressure differential is kept relatively constant, despite large changes in systemic pressure, by constriction or dilation of the efferent arteriole, as indicated in Figure 8.9.

The relationship between the important variables that describe the glomerular filtration rate can be stated approximately by a form of the Hagen-Poiseuille equation, which is a basic relation for the steady laminar flow of Newtonian fluids in straight round tubes. While it undoubtedly requires some oversimplification of the system to model glomerular filtration in this fashion, the basic relationships are elucidated by

$$Q_f = \frac{N \pi R^4 \Delta P}{8 \mu \Delta z} \tag{8.39}$$

where

Q_f = glomerular filtration rate, ml/min
N = total number of pores
R = average pore radius
Δz = average pore length
μ = plasma viscosity, 0.01 gm/cm sec
ΔP = net pressure difference (25 mm Hg)

For a glomerular filtration rate of 150 ml/min, the total number of pores consistent with the other values given for the variables may be estimated as 10^{13}. This gives a total equivalent pore cross-sectional area of about 1,000 cm^2, which is consistent with experimentally measured values. Equation 8.39 may also be used to predict the effects of pressure changes or reduction in pore area on the glomerular filtration rate.

It is convenient to simplify (8.39) by combining the variables normally constant to express the filtration rate in terms of the permeability and pressure difference as

$$Q_f = K_f \Delta P_f \tag{8.40}$$

where $K_f = N \pi R^4 / 8 \mu \Delta z$, measured in ml/min mm Hg and having a value for this situation of 6 ml/min mm Hg.

The second significant mass transfer process to occur in the nephron is the osmosis which takes place as the filtrate is convected through the proximal tubule and into the descending portion of the loop of Henle. Due to the unbalance in nondiffusible or nonpermeating solutes between the filtrate inside the tubule and the interstitial fluid being perfused by the vasa recta outside the tubules, the concentration of the solvent (in this case water) will be higher inside the tubule than outside. The resulting flow of water through the tubule wall, actually diffusion, is referred to as osmosis due to the osmotic pressure that develops and the fundamental nature of the selective membrane that permits it. Nevertheless, in computing the local flow rate of any substance, including water, we can resort to an integrated form of Fick's law, such as

$$J_i = \mathfrak{D}_i \, \Delta C_i / \Delta r \tag{8.41}$$

where

J_i = flux of any diffusible specie, meq/cm^2 min
$\mathfrak{D}_i$ = diffusivity of i, cm^2/min
ΔC_i = concentration difference of solute, meq/ml
Δr = tubule wall thickness, cm

Again it is more convenient to deal with permeabilities and flow rates rather than fluxes and diffusivities. Multiplying both sides of (8.41) by the wall area per unit length of tubule, we obtain

$$Q_i = J_i A = \mathfrak{D}_i A \, \Delta C_i / \Delta r = K_i \, \Delta C_i \tag{8.42}$$

where K_i is now the permeability (or mass transfer coefficient) for species i and Q_i is the equivalent volumetric flow rate for that species.

Since the flow of water is by far the most significant flow through the tubule wall, we can single it out as the sole osmotic process and refer to the changes in other specie concentrations as the result of diffusion. The concentration of all species inside the tube will be changed regardless of their diffusion effects, since the removal of water will make all the nondiffusible solutes more concentrated and will change the others.

The flux of water will be given the symbol Q_D and will be equated to the product of a permeability and an equivalent driving force—the overall concentration difference of all the nondiffusible j species $\overline{\Delta C_j}$, so that

$$Q_D = K_D \, \overline{\Delta C_j} \tag{8.43}$$

The concentration changes along the length of the tubule for any diffusible species may now be derived by combining (8.42), which gives the rate of solution flow at any position, with the material balance equation for convective mass transfer similar to (8.14). For any permeating solute i,

$$(R_D/2) \frac{d}{dx} (Q_f C_i) = -K_i \, \Delta C_i \tag{8.44}$$

where

R_D = tubule radius, about 25 microns
Q_f = flow rate of filtrate in the tubule
ΔC_i = concentration of any permeating solute
x = distance along the tubule (total length about 30 mm)

Since the quantity Q_f is itself decreasing with length due to loss through the walls, a balance equation is needed for that quantity. Proceeding as before,

$$(R_D/2)\,(dQ_f/dx) = -K_D\,\overline{\Delta C_j} = -Q_D \tag{8.45}$$

since the loss of water through the walls will cause a net decrease in filtrate flow through the tubule with length.

By the time the filtrate reaches the beginning of the descending loop, the flow rate has been reduced to an overall value of about 40 ml/min while the total concentration of impermeable solutes has stayed relatively constant at an osmolarity of 300 milliosmoles (mosm). In the descending portion of the loop of Henle the flow rate is further reduced to about 25 ml/min while the osmolarity rises to 1,200 mosm/liter. However, due to the countercurrent superimposition of the vasa recta, as indicated schematically in Figure 8.9, the overall concentration driving force is never greater than about 100 mosm/liter at any point along the length. Using these data, the permeability of the proximal tubule can be estimated from (8.45) as

$$K_D = R_D\,\Delta Q_f/2\,\Delta x\,\Delta C_s = 4 \times 10^{-4}\ \text{ml/min mosm} \tag{8.46}$$

while that of the descending loop is

$$K_D = R_D\,\Delta Q_f/2\,\Delta x\,\Delta C_s = 1 \times 10^{-4}\ \text{ml/min mosm} \tag{8.47}$$

To predict the concentration variation of a nondiffusible solute j, it is only necessary to make another material balance along the length to account for the effect of decreasing flow, by

$$\frac{d}{dx}(Q_f\,C_j) = 0 \tag{8.48}$$

and use (8.45) to get

$$dC_j/dx = 2\,C_j\,K_D\,\overline{\Delta C_j}/Q_f\,R_D \tag{8.49}$$

It should be noted that Q_i, the flow of solute i, may be either positive and outward from the system or negative and into the system. In either case, (8.43–8.45) and (8.49) are the basic relations for the descending tubule. Figure 8.10 summarizes the mass transfer situation for the descending portion of the loop. The values of Q_f, ΣC_i, and ΣC_j at the bottom of the loop may be computed if the permeabilities are known, and these may be compared with experimental values. The total filtrate flow decreases from an overall value of 150 ml/min to 25 ml/min during the passage through the proximal tubule and the descending loop portion, while the total osmolarity is increased from 300 mosm/liter to 1,200 mosm/liter. This implies that there is a net loss of 15 mosm of solutes per minute lost from these two portions of the nephron. Also, as we will show, there is a

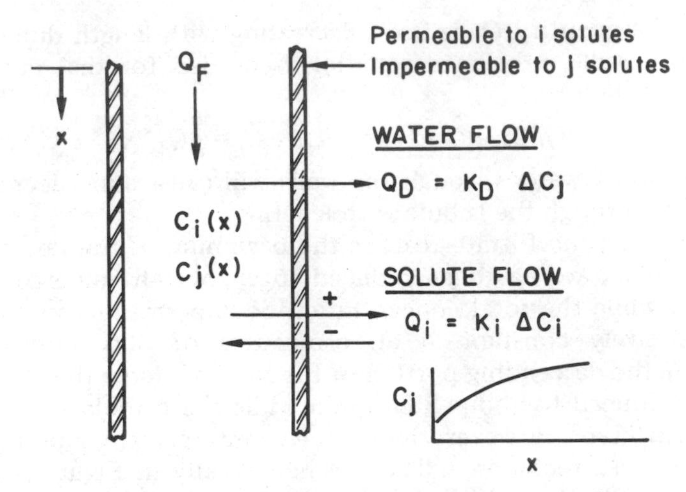

Fig. 8.10. Convective and diffusive mass transfer.

net gain of solutes during the descending portion of about 18 mosm/min. In this regard, net losses from the tubule are denoted as *reabsorption*, while net gains are denoted as *secretion*.

The computation is as follows:

Solutes entering proximal tubule
= 300 mosm/liter × 0.150 liter/min = 45 mosm/min

Solutes entering descending loop
= 300 mosm/liter × 0.040 liter/min = 12 mosm/min

Solutes leaving descending loop
= 1,200 mosm/liter × 0.025 liter/min = 30 mosm/min

Stated more formally:

$$\int_0^L (dQ_f/dx)\, dx = \int_0^L K_D \; d\,(\overline{\Delta C_j}) = 125 \text{ ml/min} \tag{8.50}$$

$$\int_0^L (dC_i/dx)\, dx = \int_0^L K_i \; d\,(\Delta C_i) = -15 \text{ mosm/liter} \tag{8.51}$$

where L = the entire length of the proximal tubule and descending loop.

It is evident from these relationships that longer loops will have more exchange capability than shorter ones, all else being equal, and that one means of varying the amount of secretion or reabsorption of water and solutes would be to vary the permeability of the tubules. Both these features have been widely studied.

Another important point here is that for a steady state to be maintained, the solution and its solutes that represent a net loss from the tubule must be picked up somewhere else—in this case in the vasa recta. Since the blood flow through this element is very much greater than the filtrate flow in the nephron (nearly 20 times more than at the bottom of the loop), the relatively small amount of solutes to be reabsorbed may be picked up without a measurable change in concentration on the blood side for the input and output streams.

Experiments have shown that as the filtrate (urine) moves through the ascending portion of the loop of Henle, the situation regarding mass transfer is quite different. It has been hypothesized that the ascending portion is impermeable to water passage, so that Q_f is constant along the length, until the distal tubule is reached. In order to explain how the urine concentration of solutes can drop in the presence of a positive concentration driving force in the vasa recta, which in turn is required to produce the high concentration observed at the bottom of the loop, it has been hypothesized that active transport takes place along the ascending portion of the loop. Solutes are transferred out of the tubule against a higher concentration in the interstitial fluid perfused by the vasa recta. While no mechanistic theory completely analogous to Fick's law may be written for this process, two conclusive quantitative statements can be made. The first is, that as in the stomach-acid example in Chapter 1, there must be a net consumption of energy by the process. The second is that the flux of solute may be adequately modeled by the expression

$$J_i^* = k_i^* C_i \tag{8.52}$$

and the flow by

$$Q_i^* = K_i^* C_i \tag{8.53}$$

where the k_i^* and the K_i^* are the active transport coefficients. The total amount of solutes transported out (reabsorbed) in the ascending loop can be estimated from available data, since osmolarities of about 100 mosm/liter have been measured near the distal tubule. The computation proceeds as follows:

Solutes entering ascending loop
= 0.025 liter/min × 1,200 mosm/liter = 30 mosm/liter
Solutes leaving ascending loop
= 0.025 liter/min × 100 mosm/liter = 2.5 mosm/liter
Net loss (reabsorption) = 27.5 mosm/liter

Using an average value for all of the C_i's, K_i^* may be estimated from (8.53) as 40 ml/min mosm.

The variation of C_i with length in the ascending loop can then be given by the material balance equation

$$dC_i/dx = -2\, K_i^*\, C_i/Q_f\, R_D \tag{8.54}$$

In the distal tubule the primary mass transfer process is once again osmotic flow of water out of the tubule coupled with diffusion of solutes into the tubule from the more concentrated surroundings. The equations that describe the proximal tubule again apply. A total of 3 mosm of solute are transferred in (secreted) to the tubule. The total flow drops to about 18 ml/min and the osmolarity rises back to 300 mosm/liter. The corresponding value for the permeability estimated by (8.46) is 1×10^{-4} ml/min mosm.

In the collecting duct the above processes continue, and in addition some active transport takes place. The overall urine flow rate is drastically reduced in the collecting duct during normal kidney function, while the solute concentrations rise appreciably. There is a net reabsorption in this section of 4 mosm/min, the flow drops to 1 ml/min, and the osmolarity rises to about 1,200 mosm/liter during normal function. Once again the permeability estimate is 1×10^{-4} ml/mosm min. Needless to say, slight changes in the permeability of this portion of the nephron can have the greatest effect on both the total amount of urine secreted and on its composition. The amount reabsorbed from the collecting duct may be given quantitatively by (8.50) as

$$Q_f = \int_0^L (dQ_f/dx)\, dx = \int_0^L K_c(\overline{\Delta C_j}) = 17 \text{ ml/min} \tag{8.55}$$

where K_c is the permeability of the duct.

Table 8.4 summarizes the values that have been used and calculated in this section and are typical of experimental values found in the literature regarding renal physiology.

This section has demonstrated how several different mass transfer mechanisms together can play important roles in the function of a

Table 8.4. Typical Mass Transfer Properties of the Kidney

Section	Q_f (ml/min)	C_i (mosm/liter)	*K* (ml/min mosm)
Glomerulus	150	300	6*
Proximal tubule	40	300	4×10^{-4}
Descending loop	25	1,200	1×10^{-4}
Ascending loop	25	100	. . .†
Distal tubule	18	300	1×10^{-4}
Collecting duct	1	1,200	1×10^{-4}

*Units are ml/min mm Hg.
†Value of active transport coefficient = 40 ml/min mosm.

single organ and has also shown how these processes can be described in a quantitative fashion.

Example 8.6. Hemodialysis

In this example we shall consider in depth the system first introduced in Example 2.2—the hemodialysis device employed as a supplement or replacement for human kidney function. In this apparatus, blood is passed extracorporeally through a device that brings it into contact with a thin cellophane membrane that in turn has its reverse side exposed to a specially prepared circulating fluid, referred to as the dialysate fluid. Figure 8.11 depicts the geometry and schematic flow encountered in a typical hemodialysis unit.

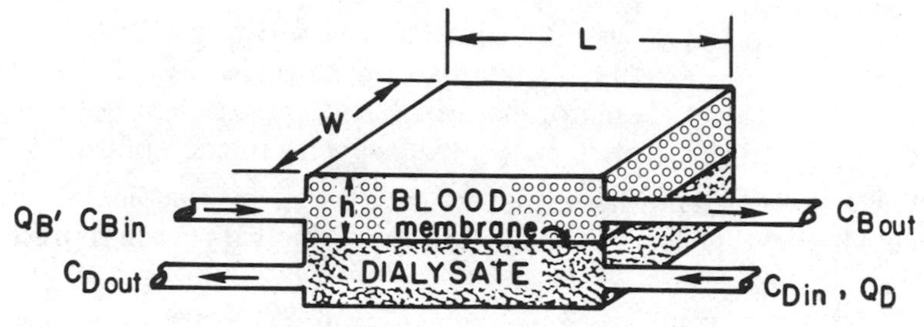

Fig. 8.11. Hemodialyzer.

The overall material balance equation for any species in this type of unit, as given in Chapter 2, is

$$Q_B\,(C_{B\ \mathrm{in}} - C_{B\ \mathrm{out}}) = Q_D\,(C_{D\ \mathrm{out}} - C_{D\ \mathrm{in}}) \tag{8.56}$$

The overall driving force for the rate of mass transfer for the species being removed will be the difference in concentrations in the blood phase and in the dialysate phase. As in many mass transfer processes the diffusing species will encounter a series of resistances in passing from one phase to the other. Figure 8.12 depicts, for a solute

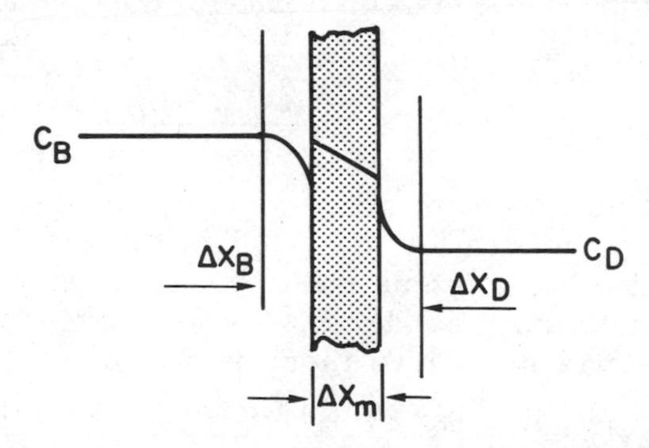

Fig. 8.12. Hemodialysis membrane.

specie, a typical concentration profile at some point along the membrane for the case where the distribution coefficient (or the partition coefficient) of the membrane is less than unity, that is, the solute is more soluble in the membrane at a given concentration than in the surrounding fluids.

As in Example 8.1, we can write an expression for the mass transfer flux across the three resistances indicated in Figure 8.12 as

$$J_i = K(C_B - C_D) \tag{8.57}$$

where

$$1/K = \Delta x_D / \mathfrak{D}_D + \Delta x_M / \phi \mathfrak{D}_m + \Delta x_B / \mathfrak{D}_B \tag{8.58}$$

where

K = overall mass transfer coefficient
Δx = film thickness for each phase
ϕ = membrane distribution coefficient
$\mathfrak{D}$ = diffusivity of the species in each phase

For the urea-cellophane membrane system in a typical countercurrent dialyzer such as the Kiil, the following values are representative:

$\phi = 0.50$ $\quad$ $\mathfrak{D}_B = \mathfrak{D}_D = 10^{-5}\ \text{cm}^2/\text{sec}$
$\Delta x_m = 0.025$ mm $\quad$ $\Delta x_B = 1$ mm
$\mathfrak{D}_m = 10^{-7}\ \text{cm}^2/\text{sec}$ $\quad$ $\Delta x_D = 4$ mm

Substitution of these values into (8.58) shows that the resistance of the fluid films are comparable in size to that of the membrane. This continues to be a major problem in dialyzer design. Specifically, the mass transfer rate continues to be limited by fluid mechanical considerations rather than by membrane properties.

When (8.57) is integrated over the entire dialyzer to produce the overall rate, assuming the overall coefficients to be independent of position, the following expression results for the overall transfer rate:

$$Q_i = J_i A = K A \frac{(C_{B\ \text{in}} - C_{D\ \text{out}}) - (C_{B\ \text{out}} - C_{D\ \text{in}})}{\ln\left[\dfrac{(C_{B\ \text{in}} - C_{D\ \text{out}})}{(C_{B\ \text{out}} - C_{D\ \text{in}})}\right]} \tag{8.59}$$

where A = the total area for transfer, WL in Figure 8.11. The last term represents the logarithmic mean concentration driving force for a countercurrently operated device. For a cocurrently operated device the appropriate mean driving force is given by

$$\Delta C_{lm} = \frac{(C_{B\ \text{in}} - C_{D\ \text{in}}) - (C_{B\ \text{out}} - C_{D\ \text{out}})}{\ln\left[\dfrac{(C_{B\ \text{in}} - C_{D\ \text{in}})}{(C_{B\ \text{out}} - C_{D\ \text{out}})}\right]} \tag{8.60}$$

and for a device such as the Kolff dialyzer, where the dialysate side is well stirred rather than flowing,

$$\Delta C_{lm} = \frac{(C_{B\ in} - C_{B\ out})}{\ln\left[\frac{(C_{B\ in} - C_D)}{(C_{B\ out} - C_D)}\right]} \qquad (8.61)$$

Equation 8.59 may be somewhat simplified for the usual case where the concentration of the removed species is very low in the dialysate fluid, so that the values of $C_{D\ in}$ and $C_{D\ out}$ are close to zero in comparison with the blood-side concentrations. Then (8.59) may be combined with the material balance equation (8.56) to produce

$$Q_i = Q_B(C_{B\ in} - C_{B\ out}) = K A \frac{(C_{B\ in} - C_{B\ out})}{\ln(C_{B\ in}/C_{B\ out})} \qquad (8.62)$$

and this in turn may be solved for $C_{B\ out}$ to produce

$$C_{B\ out} = C_{B\ in} \exp(- K A/Q_B) \qquad (8.63)$$

which can be used as a fundamental design equation for a countercurrent dialysis unit. Also, (8.63) is the "missing equation" which was needed in Example 2.2 to close the relationship between the beginning of the dialysis procedure.

With a rectangular geometry the area, channel height h, and the blood volume of the device V are related by

$$A = V/h \qquad (8.64)$$

Often it is desired to maximize the mass transfer rate, or clearance, of the artificial kidney for a given limited blood volume. Then it becomes necessary to examine the relation between the flow rate and channel height in the device, since the overall pressure drop through the device is usually also limited. Employing the Hagen-Poiseuille equation of fluid mechanics, which was used earlier in this chapter to describe glomerular filtration, this flow-pressure-volume relation can be approximated by

$$Q_B = (\pi D^4\ W h^3\ \Delta p)/(12 \mu L_c \pi D^4 + 128 \mu L W h^3) \qquad (8.65)$$

where

D = equivalent diameter of the vascular system, about 0.5 cm
L_c = equivalent length of the circulatory system, about 800 cm
p = available pressure drop for the device, about 40 mm Hg
μ = viscosity of blood, 0.02 dyne sec/cm^2

L, W, and h are as shown in Figure 8.11. Equation 8.65 may be obtained by considering the flow through the circulatory system and through the device, applying the Poiseuille equation, considering the resistances in series and equating the total flow through each part of the system.

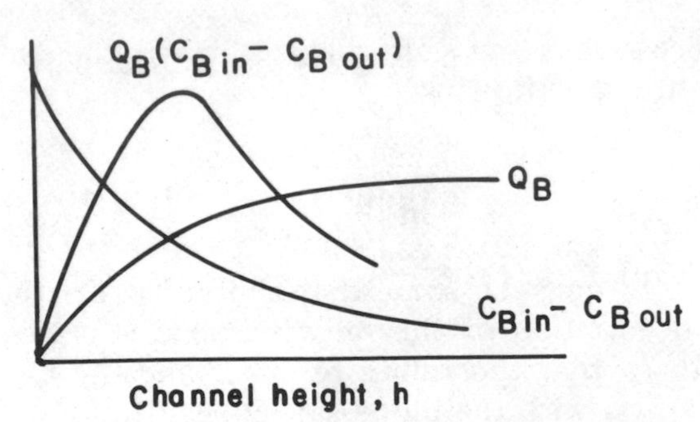

Fig. 8.13. Flow rate concentration versus channel height.

Figure 8.13 depicts a plot of the flow rate Q_B, given by (8.65), the concentration difference $C_{B\ out} - C_{B\ in}$, given by (8.63) and (8.64), and the product of these two quantities, which is equivalent to the total amount of species removed in the device, given by (8.62).

Figure 8.13 demonstrates that there is an optimal channel height h for a given mass transfer coefficient K, dialyzer volume V, and pressure drop ΔP, at which the clearance reaches a maximum due to the trade off between increasing flow rate and decreasing concentration difference.

The analysis may be extended one step further by realizing that the mass transfer coefficient itself will be a function of the flow rate Q_B, since increases in the blood flow velocity will affect the blood film thickness, Δx_B, and change that resistance accordingly. This effect may be approximated by

$$1/K = 1/K_D + 1/K_M + 1/K_B \tag{8.66}$$

and letting

$$K_B = K_{B0}\ v/v_0 = K_{B0}\ (Q_B/Q_{B0})(h_0/h) \tag{8.67}$$

then this effect may be included in the analysis. It is worth noting that the correlations discussed in Chapter 7 would indicate that $v^{0.8}$ would be a better choice for the velocity dependence, but the linear approximation offers the virtue of simplicity at this point and demonstrates the effect adequately.

When this effect is included and the results rearranged in terms of the clearance of the device ψ, where

$$\psi = Q_i/C_{B\ in} \tag{8.68}$$

for the case where $C_D = 0$, the result is then

$$\psi = \frac{\pi D^4\ V \Delta P\ h^2}{\beta}\ [1 - \exp(-\ \beta/\pi\ D^4\ \Delta P\ K\ h^3)] \tag{8.69}$$

where $\beta = 12\,\mu\, L_c^2\, \pi\, D^4 + 128\,\mu\, L\, h^2\, V$ and $K = K(h)$ as in (8.66) and (8.67).

Some typical results of these relations are shown in Figure 8.14, which gives the clearance as a function of channel height h for different channel lengths L. This figure dramatically indicates that

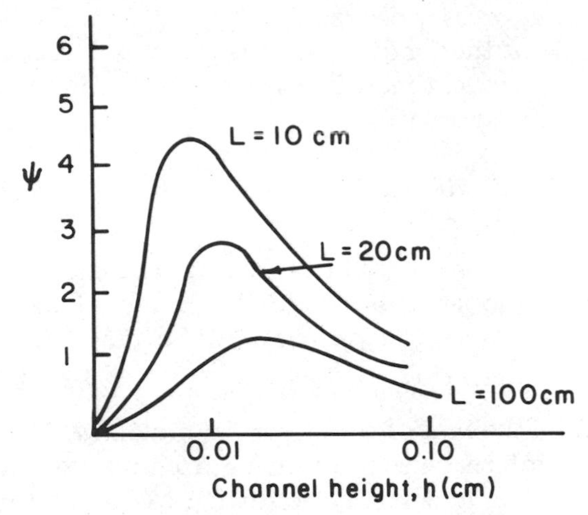

Fig. 8.14. Clearance versus channel height.

dialyzers with short narrow channels will provide the best clearance for a fixed volume and pressure drop. Since the fixed volume would in turn require a greater number of short narrow channels, this analysis then implies that the human kidney is relatively well designed —a fact already much in evidence.

This concludes a brief treatment of the principles of mass transfer as applied to the design and operation of hemodialysis devices. Obviously, extensions of these principles can lead to more general or more specific results for different designs and schemes of operation.

Example 8.7. Mass Transfer through Membranes

In this section we will consider in finer detail the processes involved in the transport of a nonelectrolyte through a membrane and attempt to relate in a fundamental fashion the important processes of osmosis, diffusion, ultrafiltration, and hydrodynamic flow. It is hoped that this section can tie together some of the concepts encountered throughout this chapter. To accomplish this, it will be necessary to consider some of the basic principles of nonequilibrium thermodynamics and also to reintroduce some concepts of equilibrium thermodynamics as they were first presented in Chapter 1.

At the outset we must again state the concept that the molar flux of a species J_i can result from gradients in other physical quantities than concentration. For instance, the mass flux that results from a temperature gradient is referred to as *thermal diffusion*, that from a pressure gradient as *pressure diffusion* (not to be confused with convection). While these effects are normally very small in systems of interest, an understanding of this cross-coupling of fluxes and driving forces is desirable.

The first law of thermodynamics, as presented in Chapter 1, may be reintroduced here and modified for the case where mass is added differentially to the system:

$$d\tilde{U} = \tilde{Q} - \tilde{W} + \sum_i \mu_i \, dn_i \tag{8.70}$$

where the new term on the right represents the energy added to the system when additional mass is added. It introduces the thermodynamic quantity of *chemical potential*, defined as

$$\mu_i = (\partial \tilde{U}_i / \partial n_i)_{\tilde{S}, \tilde{V}, n_j \text{ constant}} \tag{8.71}$$

which can also be thought of as the partial molal internal energy, that is, the change in internal energy resulting from mass addition.

The work term in (8.70) may be further expanded to read

$$-\tilde{W} = P \, d\tilde{V} + \tilde{e} \, d\psi \tag{8.72}$$

where

$\tilde{e}$ = molal charge
ψ = electrical potential

The second law and its corollary are rewritten here as

$$d\tilde{S}_{\text{system}} = \tilde{Q}/T + \tilde{R}/T \tag{8.73}$$

$$d\tilde{S} = d\tilde{S}_{\text{system}} + d\tilde{S}_{\text{surr}} \geqslant 0 \tag{8.74}$$

and the definition of *equilibrium* as

$$d\tilde{S}/dt = 0 \text{ at equilibrium} \tag{8.75}$$

The basic idea that underlies nonequilibrium thermodynamics, or the study of transport phenomena, is that the major results of equilibrium thermodynamics are still usable for small deviations away from equilibrium such as normally encountered in most transport problems. The idea is then to combine the relations given above with microscopic relations, (1) the conservation of energy equation, (2) the momentum equation, and (3) the conservation of mass equation, to produce a general relation for the rate of entropy production under conditions of nonequilibrium:

$$d\tilde{S}/dt = \sum_k J_k X_k \tag{8.76}$$

Since $\widetilde{S}$ is a maximum at equilibrium, this quantity will always be positive (or zero at equilibrium). The J_k's in (8.76) represent the various fluxes in the system (energy, momentum, mass, and so forth) while the X_k's represent the forces (gradients in temperature, velocity, concentration, and so forth) which are conjugated with these fluxes. Some examples appear in Table 8.5.

Table 8.5. Fluxes and Forces

Flux (J_k)	*Force (X_k)*
Energy q	$\Delta \ln T$
Material j_i	$\Delta \mu_i$
Momentum τ	Δv
Charge I	$\Delta\psi$
Extent of a chemical reaction R_i	Λ_i (affinity)

An interesting feature of (8.76) is that the fluxes and forces will appear naturally conjugated and are readily identifiable without further manipulation. In 1931 Onsager completed the summary of three important relations relative to these forces and fluxes. A synopsis of these might read:

1. In general, the fluxes J_k are coupled to all the forces of equal order or of order varying by an even number by

$$J_k = \sum_{j=1}^{N} L_{kj} X_j \tag{8.77}$$

   where the L_{kj}'s are called the phenomenological coefficients, examples of which are the diffusion coefficient, the viscosity, and the thermal conductivity. N refers to the total number of forces present of appropriate order.
2. The matrix of the L_{kj}'s is symmetric, that is

$$L_{kj} = L_{jk} \tag{8.78}$$

3. The dissipation function, $d\widetilde{S}/dt$, is invariant to the choice of the J's and X's (that is, to how the forces and fluxes are chosen) as long as all effects are included.

All three of these conclusions are important for the example, with which we are concerned. They can be applied to gain some insight into the case of the membrane transport of a nonelectrolyte in solution.

Figure 8.15 depicts the general situation that will be considered.

The only forces present in this system are those that can be related to the chemical potential $\widetilde{\mu}$ by

$$\widetilde{\mu}_i = \widetilde{V}_i\,\Delta P + RT\,\Delta \ln C_i + \widetilde{e}_i\,\Delta\psi \tag{8.79}$$

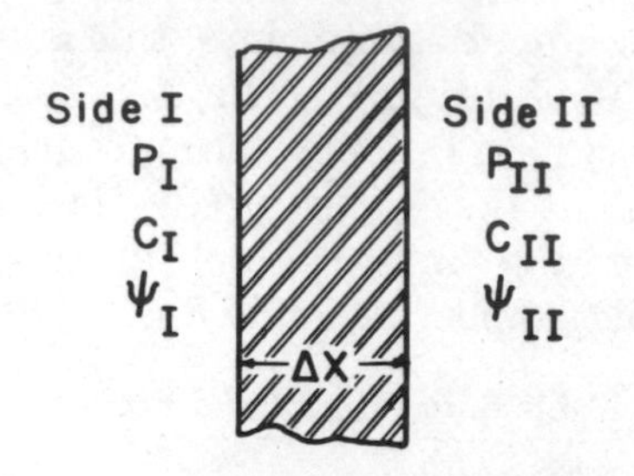

Fig. 8.15. Membrane mass transfer. C_W = concentration of water, C_S = concentration of solute.

where $\Delta = \text{I} - \text{II}$, and the i can be either W for water or S for solute. In the case we are considering $\widetilde{e}_i = 0$.

The fluxes conjugated to these forces, represented by J_k, may be written as

$$J_k = \sum_j L_{kj} \, \Delta\widetilde{\mu}_j \tag{8.80}$$

For the water

$$\Delta\widetilde{\mu}_W = \widetilde{V}_W \, \Delta P + RT \, \Delta \ln C_W \tag{8.81}$$

Since $\Delta C_W = -\Delta C_S$, and using $\Delta\pi = RT \, \Delta C_S$ (the osmotic gradient),

$$\Delta\widetilde{\mu}_W = \widetilde{V}_W \, \Delta P - (RT \, \Delta C_S / C_W) = \widetilde{V}_W \, \Delta P - \Delta\pi / C_W \tag{8.82}$$

For the solute,

$$\Delta\widetilde{\mu}_S = \widetilde{V}_S \, \Delta P + \Delta\pi / C_S \tag{8.83}$$

The flux that is conjugated to the ΔP will be labeled J_V and that to the $\Delta\pi$ as J_D. Then, since the entropy production rate is invariant to the choice of J's and X's,

$$d\widetilde{S}/dt = J_W \, \Delta\widetilde{\mu}_W + J_S \, \Delta\widetilde{\mu}_S = J_V \, \Delta P + J_D \, \Delta\pi \tag{8.84}$$

Substituting (8.82) and (8.83) into this expression and rearranging:

$$J_V = J_W \, \widetilde{V}_W + J_S \, \widetilde{V}_S \tag{8.85}$$

$$J_D = J_S / C_S - J_W / C_W \tag{8.86}$$

On close inspection we note that J_V now represents the total volumetric flow, while J_D is the net exchange flow (vol of solution/ area time). Then, expanding from (8.80).

$$J_V = L_P \, \Delta P + L_{PD} \, \Delta\pi \tag{8.87}$$

or

Total volumetric flow = Poiseuille flow + osmotic flow

and

$$J_D = L_{DP}\, \Delta P + L_D\, \Delta \pi \tag{8.88}$$

or

Net exchange flow = ultrafiltration + diffusion

From the Onsager results we also know that $L_{PD} = L_{DP}$. Also, we know from previous considerations that, remembering (8.38),

$$L_P = \pi R^4 / 8\mu\, \Delta x \tag{8.89}$$

$$L_D = \mathfrak{D}_{SW} / RT\, \Delta x \tag{8.90}$$

If the membrane is coarse and *nonselective*, then $L_{PD} = 0$. If the membrane is ideally semipermeable, then $J_S = 0$, $L_{PD} = -L_P = -L_D$. This enables us to define a *reflection coefficient* σ such that

$$\sigma = -L_{PD}/L_P, \text{ so that } 0 \leqslant \sigma \leqslant 1 \tag{8.91}$$

For an ideal semipermeable membrane $\sigma = 1$, and for a completely nonselective membrane $\sigma = 0$. Then the volumetric flow can be written as

$$J_V = L_P(\Delta P - \sigma\, \Delta \pi) = (R^4\, \pi / 8\, \mu\, \Delta x)\, (\Delta P - \sigma\, \Delta \pi) \tag{8.92}$$

In an osmometer after a long time, $J_V = 0$, and

$$\Delta P^0 = \sigma\, \Delta \pi \tag{8.93}$$

It can be seen that σ, the reflection coefficient, relates the true osmotic pressure $\Delta \pi$ to the measured osmotic pressure ΔP^0 and is a potential source of error in osmotic measurements.

If nonpermeating solutes are present in the solution, the volumetric flow can be written as

$$J_V = L_P(\Delta P - \Delta \pi_i - \sigma\, \Delta \pi) \tag{8.94}$$

where $\Delta \pi_i$ is the difference in osmotic concentration of the nonpermeating solutes.

This example can be extended further to include the effects of electrolytes. The reader is urged to consult texts such as *Membrane Transport and Metabolism* edited by Kleinzeller and Kotyk for further details. Much of the material in this example is discussed in a review article by Adam Kachalsky in that text. Other good references are discussed in the bibliography.

In this example it is hoped that the coupled mass transfer processes of osmosis, diffusion, ultrafiltration, and hydrodynamic flow have been elucidated. All four of these processes are significant, for example, in hemodialysis and play an important role in the operation of such devices. They are also important in many physiological situations.

SUMMARY

This chapter has been devoted to applying the basic principles of mass transfer as discussed in Chapter 7 to certain physiological subsystems and to hemodialysis. Obviously, the list of similar applications to be found in physiology is very long indeed, and the reader is urged to pursue for himself the multitudes of easily approachable examples in an effort to improve not only his understanding of physiology but also his understanding of mass transfer and of the transport phenomena in general. While the sophisticated reader will also be aware of a certain degree of pedagogical oversimplification in this chapter in regard to certain physiological functions, he is certainly free to extend the analysis to any level he deems appropriate. The basic principles are before him.

Bibliography

A complete list of relevant references for this chapter would be extremely long. In this bibliography we will cite the most significant works with which we are familiar, and hope that they will serve as examples of good starting points for those interested in further study of these topics.

Several excellent articles on mass transfer in living systems are contained in:

Shrier, A. L., and Kaufmann, T. G. Mass transfer in biological systems *Chemical Engineering Progress Symposium Series*, vol. 66, no. 99, American Institute of Chemical Engineers, 1970.

Further details of the role of mass transfer in the respiratory system as discussed in this chapter may be found in:

Seagrave, R. C., and Warner, H. L. Distributed-lumped parameter model of mass transfer in the respiratory system. *Chemical Engineering Progress Symposium Series*, vol. 66, no. 99, American Institute of Chemical Engineers, 1970.

The following articles are also valuable in providing background for this subject:

West, J. B. *Ventilation/Blood Flow and Gas Exchange.* Blackwell Scientific Publications, Oxford, 1967.

Crandall, E. D., and Flumerfelt, R. W. *J. Appl. Physiol.* 23 no. 3 (1967), 944-53.

Grodins, F. S., and James, G. *Ann. N.Y. Acad. Sci.* 1(1963), 852-68.

Roughton, F. J. W. *Progress in Biophysics and Biophysical Chemistry*, pp. 56-104. Pergamon, 1959.

An article which gives some background on the relation between reaction and diffusion is found as:

Keller, K. H., and Friedlander, S. K. Investigation of steady state oxygen transport in hemoglobin solution, pp. 89-95. *Chemical Engineering Progress Symposium Series*, vol. 62, no. 66, American Institute of Chemical Engineers, 1966.

A typical example of application in the circulatory system is given in:

Johnson, J., and Wilson, T. A model for capillary exchange. *Am. J. of Physiol.* 210 no. 6 (1966), 1299-1303.

The example on urea transport is derived from work initially reported as:

Bell, R. L., Curtis, K., and Babb, A. L. Simulation of the mass transfer dynamics of the patient-artificial kidney system. *Transactions of the American Society for Artificial Internal Organs* 11 (1965), 183.

Other important references for this chapter include:

Best, C. H., and Taylor, N. B. *Physiological Basis of Medical Practice.* Williams and Wilkins, 1961.

Dowben, R. J. *General Physiology.* Harper and Row, 1969.

Pitts, R. F. *Physiology of the Kidney and Body Fluids.* Year Book Medical Publishers, 1963.

Ruch, T. C., and Patton, H. D. *Physiology and Biophysics*, 19th ed. Saunders, 1965.

A review article of some use is:

Conner, E. D., and Gainer, J. L. Diffusion in biological systems. *Chemical Engineering Progress Symposium Series*, vol. 66, no. 99, American Institute of Chemical Engineers, 1970.

Supplementary references for the section on mass transfer in the kidney are the texts by Dowben and Pitts listed above, as well as:

Davson, H. *Textbook of General Physiology.* Little, Brown, 1964.

Guyton, A. C. *Textbook of Medical Physiology.* Saunders, 1966.

The material on hemodialysis, a subject of great current interest, should be supplemented by review articles such as found in:

Dedrick, R. L., Bischoff, K. B., and Leonard, E. F., eds. The artificial kidney. *Chemical Engineering Progress Symposium Series*, vol. 64, no. 84, American Institute of Chemical Engineers, 1968.

The quantitative material on hemodialysis in Chapter 8 was suggested by the original work:

Grimsrud, L., and Babb, A. L. Optimization of dialyzer design for the hemodialysis system. *Transactions of the American Society for Artificial Internal Organs* 10 (1964), 101–16.

The material on membrane mass transfer can be supplemented by the excellent treatments found in either of the two following works:

Dowben, R. J. *General Physiology*. Harper and Row, 1969.

Kleinzeller, A., and Kotyk, A., eds. *Membrane Transport and Metabolism*. Academic Press, 1961.

Additional fundamental material concerning coupled processes is developed in:

Bird, R. B., Stewart, W. E., and Lightfoot, E. N. *Transport Phenomena*. Wiley, 1960.

Index

N

O

P

R

S

T